Reviews of

101 Physiology, Biochemistry and Pharmacology

With 22 Figures

Springer-Verlag
Berlin Heidelberg GmbH 1984

ISBN 978-3-662-31038-0 ISBN 978-3-540-39040-4 (eBook)
DOI 10.1007/978-3-540-39040-4

Library of Congress-Catalog-Card Number 74-3674

Originally published by Springer-Verlag Berlin Heidelberg New York Tokyo in 1984
Softcover reprint of the hardcover 1st edition 1984

2127/3130-543210

Contents

Indexed in Current Contents

Rev. Physiol. Biochem. Pharmacol., Vol. 101

The Mammalian Muscle Spindle and Its Central Control

MANUEL HULLIGER

Contents

Brain Research Institute, University of Zürich, August-Forel-Straße 1, CH–8029 Zürich, Switzerland

Received for publication 26 September 1983

1 Introduction

Research on the muscle spindle has been concerned with a wide range of aspects, including morphology and receptor mechanisms, the functional properties and central actions of spindle afferent neurones, and the central control and peripheral actions of fusimotor neurones. This article focuses on the properties of spindle afferents and efferents and on their role during natural movement. Morphology is considered mainly in its historic context and where it aids the understanding of function; the receptor mechanisms and the central actions of spindle afferents are dealt with but in passing.

The muscle spindle is not a simple stretch receptor organ. Instead, its afferent output is determined by the complex interaction of two types of time-dependent inputs: muscle length and fusimotor drive. The latter is provided by the motor supply to the spindle which is largely mediated by γ-motoneurones, whose sole function is fusimotion. This was established by the pioneering work of *Leksell* (1945) and *Hunt* and *Kuffler* (1951a; *Kuffler* et al. 1951) and has never since been challenged. Yet fusimotion is not solely provided by γ-motoneurones. More recently it has been appreciated that a significant fraction of the spindle's motor supply is derived from mixed skeletofusimotor (β-)motoneurones (see *Laporte* et al. 1981). These, once they are recruited, induce a measure of coupling between overall fusimotor drive and skeletomotor activity.

Fusimotor action is not uniform. Whilst all fusimotor fibres exert a biasing action on spindle afferents and so extend their working range in terms of muscle length, their effects on dynamic sensitivity fall broadly into two categories, static and dynamic, according to whether the dynamic response to muscle stretch is decreased or increased (*Matthews* 1962). Moreover, a wide potential for central actions on fusimotor neurones has been revealed over the years, so that a number of strategies with varying degrees of independence in the control of α-motoneurones and static and dynamic γ-motoneurones are easily conceived. Most influential, and repeatedly addressed in this article, were the following concepts. *Hunt* and *Kuffler* (1951a) suggested that fusimotor action might serve to maintain spindle afferent firing during shortening contractions to provide ongoing sensory feedback. Wisely, perhaps, these authors refrained from specifying particular patterns of time course of fusimotor outflow. This theme was taken up by *Merton* (1951, 1953), who proposed that γ-motoneurones were crucially involved both in the initiation and subsequent control of voluntary movements by conveying motor commands to α-motoneurones indirectly via γ-Ia loop (follow-up servo hypothesis). This possibility can now safely be dismissed, partly owing to *Vallbo's* (1971) demonstration that during voluntary contraction in man the onset of spindle afferent discharge followed rather than preceded the onset of electromyographic (EMG) activity, but also since it was shown that the execution of sterotyped rhythmic movements did not depend on the presence of afferent feedback (see *Grillner* 1981, locomotion; *Goodwin* and *Luschei* 1974, mastication; *Sears* 1964; *Eklund* et al. 1964, respiration).

Since then the concept of α-γ linkage (*Granit* 1955), envisaging an often tight coupling between skeletomotor and fusimotor activity, has acquired an increasing following. However, the advocates of α-γ linkage (or of its looser version, α-γ coactivation), whilst elevating it to a general principle of motor control (e.g. *Granit* 1970, 1979), have consistently failed to take into account the existence of separate static and dynamic fusimotor neurones (*Matthews* 1964, 1972). This is remarkable, since this distinction derived its justification not only from the observation of characteristically different peripheral actions but also from increasingly compelling evidence for a selective central control of these two types of fusimotor neurone (see *Matthews* 1972; Sect. 4.2).

Further concepts of fusimotor function are discussed in the context of this article, and a brief overview is presented in the concluding section.

2 Structure and Innervation of the Mammalian Spindle

Amongst peripheral receptors the muscle spindle is unique because of its efferent fusimotor innervation, which exposes it to direct control by the central nervous system. This has made it an attractive, intriguing object of scientific curiosity, both for those interested in receptor properties and for those concerned with higher nervous function. The historic developments have been recounted on several occasions (see, e.g. *Matthews* 1964, 1972, 1981a,b; *Barker* 1974a; *Hunt* 1974; *Boyd* 1981a), and these reviews may be consulted for further detail. Here only the most important steps will be outlined to provide a historic basis for the present picture, which enjoys an unprecedented degree of general support. The emphasis of this section is on spindle structure, yet certain functional considerations had also to be included. Moreover, a topical rather than a strictly chronological form of presentation has been chosen.

2.1 Historic Landmarks

2.1.1 The Classical Picture

During this century the main conclusions of *Sherrington* (1894) and *Ruffini* (1898) concerning structure and function of the mammalian neuromuscular spindle have been progressively refined rather than seriously challenged. Their work settled a long controversy as to wether muscle spindles were 'muscle buds', concerned with the generation of skeletal muscle fibres, or perhaps even foci of inflammatory processes (see *Matthews* 1972, 1981a). The unexpected outcome, now common knowledge, was that muscle spindles were sensory organs (*Sherrington* 1894), which were richly supplied by afferent and efferent nerve fibres (*Ruffini* 1898).

Ruffini's classical drawing (Fig. 1) shows the spindle as a fine, elongated organ consisting of small, so-called intrafusal muscle fibres which in their central portion are enclosed by a capsule formed of connective tissue. The

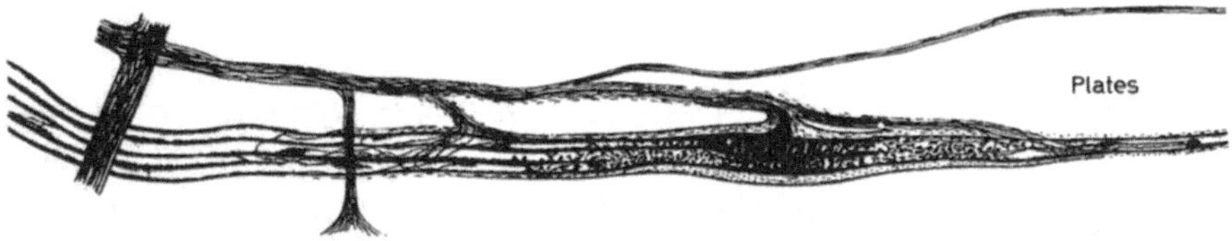

Fig. 1. The classical picture of the muscle spindle in the cat, as seen by *Ruffini* in 1898 (Fig. 1). Illustration of the central encapsulated segment with the midequatorial region (primary ending) and the juxtaequatorial regions (secondary endings). The 'motor bundle' was described by *Ruffini* as supplying axons to the skeletal muscle fibres in the vicinity of the spindle. Intrafusal muscle fibres are discernible as fine, elongated structures in the left pole of the spindle. (From *Ruffini* 1898, with permission)

capsular swelling gives the organ its distinguishing fusiform shape and also gave it its name. Moreover, and more importantly, it contains the 'sensory region' with the abundant ramifications and terminations of the afferent nerve fibres. The primary endings, which may now be taken to connect to Ia afferent fibres (see Sect. 2.1.2 and *Matthews* 1972; *Barker* 1974a, reviews), are wrapped around the intrafusal fibres as these pass through the central or 'equatorial region' (Fig. 1). The secondary endings (the terminations of group II afferent fibres) may be seen to occupy the adjacent 'juxtaequatorial' regions on either side of the primary terminals (for terminology, see Fig. 3).

Ruffini also described a third type of ending, situated on the more peripheral polar segments of the intrafusal muscle fibres, which he identified – correctly – as motor 'plate' (Fig. 1). In passing, *Ruffini* should also be given credit for illustrating a feature which has long gone unnoticed during subsequent research on the motor supply of the spindle: the nerve branch supplying the spindle in Fig. 1 also sends a "motor nerve bundle.... which goes to end exclusively in the motor end plate in the muscle fibres round about the neuromuscular spindles" (*Ruffini* 1898). We now know that the motor supply to the intrafusal fibres is not only derived from the small γ-motoneurones but that a substantial contribution is also made by the mixed skeletofusimotor or β-motoneurones, which in addition innervate extrafusal motor units (see Sect. 2.1.5). *Ruffini's* drawing cannot, of course, be taken as conclusive evidence for the presence of β-innervation, since he did not attempt to trace single motor axons. Nevertheless, the close vicinity of the skeletomotor bundle and the adjacent nerve twig to the spindle of Fig. 1, which supplies plate endings in the left spindle pole, suggests that they might indeed contain branches of individual axons innervating both extra- and intrafusal muscle fibres.

2.1.2 Sensory Nerve Endings and γ-Motoneurones

After the pioneering work of *Sherrington* and *Ruffini,* it took a long time until distinguishing functional properties could be ascribed to the three types of nerve endings of the spindle. Following the first recordings from single muscle stretch receptor afferents (*Adrian* and *Zotterman* 1926), *B.H.C. Mattews* was the first to record the electrical impulse activity from single *spindle afferent units* (*Matthews* 1931, frog; *Matthews* 1933, cat). *Matthews'* (1933) study of mammalian spindles also provided the first demonstration of *fusimotor effects,* i.e. of the excitation of spindle afferents on supramaximal stimulation of the .motor nerve to their parent muscle. Although both *Ruffini's* (1898) and *B.H.C. Matthews'* (1933) investigations indicated that the bulk of the fusimotor innervation of the spindle was mediated by small motoneurones, the identification of these motoneurones in terms of fibre size had to await the advent of more refined techniques.

The elucidation of the *fusimotor control* of the muscle spindle began with the work of *Leksell* (1945) and *Hunt* and *Kuffler* (1951a,b; *Kuffler* et al. 1951). These authors showed that spindle afferents could be excited by selective stimulation of small γ-motoneurones. These were identified by *Kuffler* et al. (1951) by their comparatively slow axonal conduction velocity (15–55 m/s; α-motoneurones: 50–120 m/s). *Hunt* and *Kuffler* introduced a powerful double single-unit preparation, which enabled them to record from single afferent fibres whilst simultaneously stimulating single efferent fibres. Using this technique they established the important and until now unchallenged fact that fusimotion was the exclusive function of γ-motoneurones, since tetanic stimulation of these fibres provoked a striking excitation of spindle afferents without causing any measurable tension in the parent muscle (*Hunt* and *Kuffler* 1951a,b; *Kuffler* and *Hunt* 1952).

In 1961 *Cooper* opened the research on the functional properties of *primary and secondary* receptor *endings.* She classified functionally single spindle afferent fibres as group Ia or group II fibres on the basis of their axonal conduction velocity. Her work clearly demonstrated that group Ia fibres were more sensitive to the velocity of muscle stretch than group II fibres. Later work has amply confirmed this distinction between the two types of afferents in terms of their dynamic sensitivity to muscle stretching (see Sect. 3.1.1 and *Matthews* 1972). Spindle Ia afferents have since been equated with the primary *afferent,* i.e. with the fibres supplying the *primary endings* (see *Matthews* 1972). Correspondingly, spindle group II afferents have been identified with the *secondary afferents,* supplying the *secondary endings* of the receptor. This rests on two kinds of histological findings: first, that the primary and secondary endings connected to intramuscular afferent fibre branches of large and small diameter, respectively; second, that the fibre diameter spectra of the intramuscular branches and of the axons in the muscle nerve showed comparable multiple peaks so that corresponding peaks could reasonably be ascribed to one and the same fibre type (see *Lloyd* and *Chang* 1948; *Rexed* and *Therman* 1948; *Hunt* 1954; *Adal* and *Barker* 1962; *Barker* 1962b; *Boyd* and *Davey* 1968). However, it is easily forgotten that the identification of afferent units in terms of their receptor endings is only the simplest and most likely interpretation of the available data. For up until now, nobody has managed, and presumably not even attempted, to trace the axons of individual receptor terminals all along their course to the recording site to show that their distal and proximal axon diameters indeed belonged to the same category.

2.1.3 Intrafusal Muscle Fibres and the Functional Subdivision of γ-Motoneurones

Following the demonstration of the fusimotor role of the γ-fibres, their targets (the intrafusal muscle fibres) came under closer scrutiny. At about the same time *Boyd* (1956, 1962), working on the cat, and *Cooper* and *Daniel* (1956, 1963), investigating human spindles, reached the conclusion that the morphological variety of intrafusal muscle fibres, which *Sherrington* (1894) had already noticed, was due to the existence of two different types of fibre. These were differentiated according to the arrangement of the muscle fibre nuclei (see Fig. 2) in the equatorial region of the spindle, and they came to be known as the large *nuclear bag fibre* and the smaller *nuclear chain fibre*. At the same time further histological investigations in several laboratories led to the view that the bewildering variety of motor endings (cf. *Ruffini* 1897, 1898) could also be reduced to two main categories (*Barker* 1962b; *Boyd* 1962), although the precise definitions of these two categories were apparently not the same in the two main camps involved (see *Barker* 1962a).

In view of such bimodality of morphological features it was compelling for neurophysiologists to look for functional equivalents. A first indication of functional dichotomy came from the observation of *Jansen* and *Matthews* (1962) that static and dynamic sensitivity of primary spindle afferents varied independently of each other during reflex-induced or spontaneous variation of overall fusimotor drive. *Matthews* (1962) then described individual γ-fibres as being either static or dynamic, according to whether they decreased or increased the size of the dynamic response of primary afferents to ramp stretches. The validity of this subdivision was later amply confirmed (e.g. *Crowe* and *Matthews* 1964a,b; *Brown* et al. 1965; *Appelberg* et al. 1966; *Bessou* et al. 1966; *Matthews* 1972, review), but not without some restriction (see below).

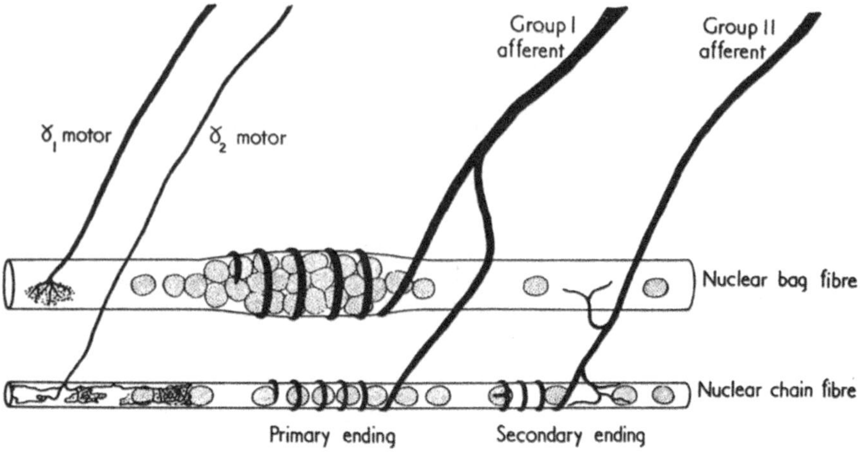

Fig. 2. Schematic view of the sensory and motor innervation of the mammalian muscle spindle, as proposed by *Boyd* (1962) and *Matthews* (1964) to encompass the main morphological and functional dichotomies that were recognized at the time: γ_1- (presumed dynamic) and γ_2- (presumed static) fusimotor axons, acting selectively on the large and thick nuclear bag fibres and the small and thin nuclear chain fibres; the primary afferent, originating from either type of intrafusal muscle fibre, as opposed to the secondary afferent, which originates largely but not exclusively from the nuclear chain fibres. (From *Matthews* 1964, with permission)

At the time, it was tempting to assemble this threefold dichotomy into a unifying model. The main feature of the scheme proposed by *Crowe* and *Matthews* (1964a) and *Matthews* (1964) (see Fig. 2) was the independent and selective control of nuclear bag and nuclear chain fibres by the two types of fusimotor fibre. Thus, dynamic γ-axons would selectively innervate nuclear bag fibres, perhaps via plate endings, which at that time were thought to occur largely (*Boyd* 1962), but not exclusively (*Barker* 1962b), on bag fibres. In contrast, static γ-axons would selectively operate chain fibres, perhaps via the second type of motor terminal, which according to *Boyd* (1962; 'diffuse γ-network') was restricted to chain fibres.

However, this scheme (Fig. 2) was challenged from the outset. *Barker* emphasized that the motor innervation of the spindle by the different types of γ-fibres was non-selective (*Barker* 1962b; *Barker* and *Ip* 1965), since he found that individual γ-fibres could terminate with plate endings either on bag or on chain fibres or, most seriously, on both. The same was found for those γ-fibres which supplied the second type of motor ending, which eventually was named 'trail ending', following *Barker's* terminology (1962b). Thus at that time the possibility of a non-specific motor innervation of the spindle (often referred to as 'cross-innervation') already was a major issue. On the other hand the scheme's dichotomies as such were not as yet challenged.

More recent research has shown that the morphological categories were oversimplifications, whilst, remarkably, the functional subdivision has stood the test of time more successfully. First, the dichotomic classification of the motor endings came under attack when *Barker* and his coworkers reached the conclusion that there were three types of endings, and that on light-and electron-microscopic grounds the plate endings had to be subdivided into p_1 and p_2 plates (*Barker* 1966; *Barker* et al. 1970). Their differential speeds of degeneration further indicated that p_1 plates were supplied by β-axons, leaving p_2 and trail endings to be innervated by γ-motoneurones. This again was at variance with the specificity of *Matthews'* scheme, since at the time β-axons were thought to exert exclusively dynamic fusimotor action. Yet *Barker* et al. (1970) found p_1 plates also on chain fibres. This on its own might not have sufficed to bring down the scheme of Fig. 2, since much of the controversy was centred around the structure of the motor terminals, whose functional significance (in terms of static vs dynamic action) was difficult to assess.

The final blow came with the recognition of a third type of intrafusal muscle fibre when the histochemical techniques which had been widely used for the characterization of skeletal muscle fibres (see *Ariano* et al. 1973; *Burke* and *Edgerton* 1975, review) were brought to bear on intrafusal fibres. It became apparent that nuclear bag, but not nuclear chain, fibres were highly non-uniform in their enzyme profiles (*Barker* and *Stacey* 1970; *Banks* 1971; *Ovalle* and *Smith* 1972). Amongst the bag fibres there were considerable differences in the levels of activity, e.g. of oxidative enzymes and of myofibrillar ATPase. Particularly the latter appeared to be more than a mere histological marker. Instead, this finding seemed relevant functionally, since ATPase activity in muscle tissue had previously been shown to be related to the speed of contraction (*Barany* 1967). This gained further weight when it emerged that the histochemical variety was paralleled with considerable ultrastructural variety and especially when it was shown that these two sets of parameters were correlated in individual fibres (see *Banks* et al. 1977b). By 1974, agreement was reached that yet another subdivision, this time concerning nuclear bag fibres, was necessary (see *Barker* 1974b), yet it took some time until an unambiguous terminology came into common use. From 1976 onwards (see *Homma* 1976), *bag*$_1$ (or dynamic bag$_1$) fibres were differentiated from *bag*$_2$ (or static bag$_2$) fibres. The bag$_2$ fibre revealed somewhat unorthodox characteristics, since in the light microscope it showed the classical features of bag fibres even more clearly than the bag$_1$ fibre, whereas its histochemical, ultrastructural and contraction properties were much closer to those of the chain fibres (see Sect. 2.2.1.2 and *Boyd* 1981a).

In 1976 the attributes 'dynamic' for the bag$_1$ and 'static' for the bag$_2$ fibre were still controversial. The debate was centred around the old question whether the different types of intrafusal fibre were selectively innervated by static and dynamic fusimotor neurones.

2.1.4 Specificity of Static and Dynamic Fusimotor Innervation

With the recognition of static and dynamic fusimotor neurones as distinct classes, the question arose immediately how these neurones provoked their contrasting effects on the dynamic sensitivity of primary afferents. There was no shortage of possibilities.

Like any other sensory receptor organ, the muscle spindle may be subdivided into three compartments: the (mechanical) filter (essentially the intrafusal muscle fibres and any elastic or muscle tissue in series), the transducer, and the encoder (or pace-maker). Theoretically, the fusimotor fibres could exert their specific effects by acting selectively on any of these three compartments, provided that the particular compartment concerned existed in two forms, one with static and the other with dynamic properties. The most popular hypotheses on the mechanisms of static and dynamic fusimotor action have been in terms of *mechanical filter* properties. On this view each type of fusimotor neurone is thought to act upon and to specifically alter the filter properties of selected intrafusal fibres (see Sect. 3.2.3).

However, at present it cannot be excluded that the specificity of fusimotor action arises from the selection of specific (i.e. static or dynamic) *transducers* or *encoders*, each associated with particular intrafusal fibres. In such an arrangement different encoders situated on different Ia terminals would have to interact as a switching pace-maker system by competing for dominance in the final discharge pattern of the afferent axon (see *Crowe* and *Matthews* 1964b; *Lennerstrand* 1968b; *Hulliger* et al. 1977c; *Banks* et al. 1977a, 1982; *Hulliger* and *Noth* 1979; *Quick* et al. 1980). The selection of the dominant pace-maker would then be made on the basis of the highest instantaneous rate of discharge, and this dominance would be further enhanced by antidromic invasion and resetting of the other pacemakers (*Eagles* and *Purple* 1974). Finally, each type of fusimotor neurone would then merely have to bring into prominence its private and specific transducer and/or pace-maker by elevating the discharge rate of the latter. Common to all these possibilities is the element of a selective innervation of different intrafusal fibres by the two types of fusimotor neurones (as enshrined in the model of Fig. 2).

It took a long time to settle the question of specificity of innervation. The main difficulty lay in the very nature of the task, which was to correlate phenomena of a different kind which at the outset were studied in different preparations and with different techniques. The static and dynamic neurones were defined by their action on the sensitivity of primary afferents, without prejudice of whether this involved bag or chain fibres. The morphological classifications, on the other hand, were made without prior knowledge of the functional properties of the intrafusal or motor fibres seen in the microscope. It was only when the functional and morphological techniques were combined in one and the same experiment – and often on one and the same spindle – that a solution came within reach.

At the beginning there was confusion. Reports based on four different techniques all agreed that static γ-fibres innervated both bag and chain fibres non-selectively, whereas dynamic axons were specific for bag fibres (see *Laporte* and *Emonet-Dénand* 1973). The techniques used were: first, chronic denervation experiments sparing only a single static γ-fibre (*Barker* et al. 1973); second, the first application by *Brown* and *Butler* (1973, 1975) of the glycogen-depletion technique (*Edström* and *Kugelberg* 1968) to mark the intrafusal muscle fibres, operated by identified γ-axons; third, the visual observation (in isolated spindles) of intrafusal movements provoked by individual γ-fibres (*Bessou* and *Pagès* 1975); and fourth, the investigation of the intrafusal targets of single γ-axons using intracellular recording from muscle fibres and subsequent histological identification (*Barker* et al. 1972, 1978).

Nevertheless, a hint of specificity remained: in these and other studies on isolated spindles a single intrafusal fibre has never been reported to be operated by both a static and a dynamic fibre (*Bessou* and *Pagès* 1975; *Boyd* et al. 1977).

A skeptical but perfectly rational interpretation of these data would be that the morphological features used to distinguish between nuclear bag and nuclear chain fibres are simply epiphenomena and quite irrelevant for the understanding of the mechanisms of static and dynamic fusimotor action. These might be due to literally invisible differences of the filter and/or transducer properties. In support of this possibility it may be noted that in the rabbit, at least on light-microscopic evidence, all intrafusal fibres are of the bag type (*Barker* and *Hunt* 1964). Nevertheless, they are innervated by perfectly normal static and dynamic fibres, identified functionally (*Emonet-Dénand* et al. 1966; *Emonet-Dénand* and *Laporte* 1969).

The scene changed again when the bag_2 fibre gained respectability. Those who still favoured specificity proposed that all would be well if static action were restricted to the bag_2 fibre, leaving the bag_1 fibre to dynamic fusimotor neurones. *Brown* and *Butler* (1975) and *Boyd* et al. (1975) were the first to voice this possibility, and further evidence in its support soon became available from combined functional-anatomical studies on isolated spindles (*Boyd* 1976a,b; *Boyd* et al. 1977; *Barker* et al. 1978; *Banks* et al. 1978). However, those working with the glycogen-depletion method kept on supporting non-specificity, since they consistently found not only nuclear chain and bag_2 fibres but also bag_1 fibres that were depleted after stimulation of identified static fusimotor fibres (*Barker* et al. 1976a,b; *Emonet-Dénand* et al. 1980b). The dynamic axons, on the other hand, continued to be non-controversial, since they always selectively innervated nuclear bag_1 fibres only.

Thus, at this stage the controversy was about the presence or absence of static fusimotor innervation of the bag_1 fibre and about the validity of the methods used. Perhaps, glycogen depletion of bag_1 fibres was an artifact (occurring in the absence of junctional activation). Alternatively, weak static-induced contractions of bag_1 fibres might have escaped visual detection in isolated spindle preparations.

The issue has recently been settled in favour of static selectivity. The evidence is of two kinds. First, *Gladden* (1981), recording intracellularly from bag_1 fibres, found that these were readily activated by dynamic, but never by static, axons. Second, and perhaps more decisive, *Barker* and collaborators (*Banks* et al. 1981a; *Barker* and *Stacey* 1981; *Banks* 1981), working with serially sectioned and reconstructed as well as with teased spindle preparations, eventually confirmed that static γ-innervation of bag_1 fibres indeed was rare. Part of their argument rested on the assumption that trail endings are specific and exclusive markers of static γ-axons. There can be little doubt that static axons indeed preferentially supply trail endings, yet the exclusiveness of this association is more difficult to ascertain (see Sect. 2.2.1.4 and *Arbuthnott* et al. 1982).

Thus, the muscle spindle controversy (*Boyd* 1981a) has come to a temporary end. The latest verdict is that the glycogen-depletion method is liable to give artifacts (*Barker* and *Stacey* 1981), especially when spindles are 'fatigued' following long lasting activation of static γ-efferents (*Decorte* et al. 1984). The only reservations must be that so far it is not evident just why this should happen and why only bag_1 fibres should be affected. One possibility is that the prolonged regime (several hours) of ischaemia and intermittent fusimotor stimulation might cause considerable accumulation of potassium in the capsular lymph space. This could in turn depolarize and directly activate the bag_1 fibre (cf. *Kidd* et al. 1971). Alternatively, bag_1 fibres, being particularly sensitive to acetylcholine (*Gladden* 1976), might be preferentially activated by this transmitter if it leaked (under ischaemic conditions) from neighbouring motor terminals (*Arbuthnott* et al. 1982). Or else, the bag_1 fibre might be subjected to some kind of stretch activation (see *Pringle* 1949, 1978; *Hulliger* et al. 1977a; *Poppele* and *Quick* 1981), since during the intermittent activation of the other intrafusal fibres it might be exposed to continual unloading and reextension.

Comparing old with new (i.e. Figs. 2 and 3), one arrives at two main conclusions which, in view of the great mass of experimental detail, are surprisingly simple. First, the perhaps most essential feature of Fig. 2 has been rehabilitated: static and dynamic fusimotor neurones achieve their distinguishing effects by selectively operating different types of intrafusal fibres. Second, the main new features of Fig. 3 are that the nuclear bag fibres have been subdivided into two categories and that the bag_2 fibre has acquired a new master and is now under static fusimotor control.

The selectivity so far described for γ-motoneurones also applies for the other type of fusimotor neurones, the β-motoneurones.

2.1.5 β-Innervation

'β-Innervation' is a misleading term (see *Matthews* 1981a) standing for a widespread and significant phenomenon. β-Motoneurones have conduction velocities in an α-range, but their action is both skeletomotor and fusimotor. In lower vertebrates such mixed skeletofusimotor innervation is the rule (*Katz* 1949; *Gray* 1957; *Eyzaguirre* 1957, 1958). In mammals β-innervation was long held to be infrequent and perhaps a mere relic of evolution. However, mainly due to the work of *Emonet-Dênand* and *Laporte* and their collaborators, the past decade has seen a 'rise of the β-system' (*Matthews* 1981a). Yet this applies only for its frequency of occurrence and still does not permit firm conclusions about its phylogenetic position. The following treatment is not exhaustive, since β-innervation today is a non-controversial issue. Moreover, the subject has repeatedly been reviewed (*Laporte* and *Emonet-Dênand* 1976; *Laporte* 1979; *Emonet-Dénand* et al. 1980a; *Laporte* et al. 1981; *Matthews* 1981a,b).

The first convincing evidence for the existence of β-innervation in the cat came at about the same time from physiological (*Bessou* et al. 1963, 1965) and morphological (*Adal* and *Barker* 1965) investigations (both in the deep lumbrical muscles of the cat hindlimb) but after the first description in human muscles of branching motor axons supplying both extrafusal and intrafusal muscle fibres (*Cooper* and *Daniel* 1963). The small lumbrical muscles were a particularly suitable object of study, given their small numbers of spindles and motor units. Thus the anatomists succeeded in tracing branches of individual α-axons to their dual destinations inside and outside the spindle, and the physiologists only had to cope with a practicable number of combinations of afferents and efferents that had to be tested for fusimotor effects. In order to ascertain the genuine fusimotor nature of excitatory effects, the muscle had to be critically curarized to exclude extrafusal effects that might have caused fusimotor-like excitation of spindle afferents. The first β-fibres that were so demonstrated were slowly conducting axons in the α-range. Invariably, they exerted dynamic fusimotor effects. In retrospect this is probably more surprising than it was at the time, since the class of fast-conducting α-fibres has since also been shown to contain mixed skeletofusimotor neurones (see below).

In larger muscles β-axons were not often encountered merely by chance. This led to the notion that β-axons were a speciality of small muscles and that they were invariably dynamic. This view was first challenged by *Barker* et al. (1970), who identified β-axons by their p_1 terminals (see above, Sect. 2.1.3). and who found that p_1 plates indeed were very common. They suggested therefore that β-innervation should be part and parcel of the motor inventory of most muscles.

This proposal could, however, only be put to a functional test after improvements in experimental technique. First, the pharmacological method of critically blocking skeletomotor contraction was replaced by a more convenient fatigue-block technique, relying on prolonged high-frequency stimulation (*Emonet-Dênand* and *Laporte* 1974). This interferes relatively selectively with the generation of extrafusal tension, without significantly affecting the excitatory fusimotor effect (but see *Matthews* 1981b; *Jami* et al. 1982, and below). This finding is interesting in its own right, for considerations of motor control during fatigue. Second, the testing task was cut down to manageable size by partially denervating larger muscles and restricting the analysis to the afferents and efferents coursing in a small natural branch of the motor nerve. If anything, this should have led to underestimates of the true incidence of β-innervation, since fusimotor effects from efferent axons taking another route to the muscle would no longer have been detected.

β-Innervation has now been found in a number of different muscles and species (cat, monkey, rat, rabbit, man; *Cooper* and *Daniel* 1963; *Bessou* et al. 1963, 1965; *Adal* and *Barker* 1965; *Emonet-Dênand* et al. 1966, 1970, 1975; *Barker* et al. 1970; *Swash* and *Fox* 1972; *Andrew* and *Part* 1974; *Barker* 1974a; *Emonet-Dênand* and *Laporte* 1975; *McWilliam* 1975; *Laporte* and *Emonet-Dênand* 1976; *Cheney* and *Preston* 1976b; *Harker* et al. 1977; *Murthy* et al. 1982; *Murthy* 1983).

Moreover, it has been recognized that β-fusimotor action is not exclusively dynamic. There is now good evidence that fast β-axons (above 85 m/s) exert static fusimotor action. Frequently these axons innervate skeletal muscle fibres of the fast, oxidative-glycolytic type. In contrast, the slower β-axons (below 85 m/s), which mostly supply muscle fibres of the slow oxidative type, exert dynamic action (*Barker* et al. 1977; *Harker* et al. 1977; *Jami* et al. 1978, 1979; *Laporte* 1979; *Emonet-Dênand* et al. 1980a; *Jami* et al. 1982; *Murthy* 1983).

The reasons for the late recognition of the static β-axons were largely technical. The crux of the matter is that the two methods routinely employed (critical curarization, *Bessou* et al. 1965; critical fatiguing, *Emonet-Dênand* and *Laporte* 1974) are more discriminative (between extra- and intrafusal junctions) for slow than for fast motor axons (*Emonet-Dênand* and *Laporte,* personal communication; see also *Jami* et al. 1982). The breakthrough eventually came when glycogen-depletion studies demonstrated that fast β-action usually involved chain fibres (see references above). For these fibres there is now little doubt concerning their role as specific effectors of static fusimotor action (see Sects. 2.1.4 and 2.2.1.4). With the existence of static β-axons firmly established, it has become easier to detect them relying on a number of tests which reflect their distinguishing properties, rather than by using a single criterion (see *Jami* et al. 1982).

The intrafusal target fibres of β-axons are not as yet investigated in the same detail as is the case for γ-fibres. Only a few observations on isolated spindles have so far been reported, so that the present evidence for β-fibres is entirely based on glycogen-depletion studies and therefore qualified by the general reservations concerning this method (see Sect. 2.1.4). Nevertheless, it appears that fast, static β-axons act preferentially on chain fibres, especially on the longest amongst them (*Jami* et al. 1978), whilst the slow, dynamic β-axons operate almost exclusively nuclear bag_1 fibres (*Barker* et al. 1977, 1980).

2.2 Current Picture in the Cat

2.2.1 Overview

In this section the current picture of the cat muscle spindle is presented in coherent form with the emphasis on established facts. Comprehensive listing of references is mostly omitted for the sake of a more fluent presentation. The most relevant references are appended at the end of each paragraph to give, together with those of the preceding sections, an adequate coverage of the field. The main features are summarized in the schematic drawing of Fig. 3 and Tables 1 and 2 (taken from *Boyd* 1981a). Minor issues and open questions are dealt with in the small print sections.

2.2.1.1 Muscle Spindle. The muscle spindle is a sense organ which has been found in widely varying numbers in virtually all striated muscles of the cat, the extraocular muscles being a notable exception (in cat but not in man). It consists of (intrafusal) muscle fibres whose central segment is enclosed by a capsule of connective tissue, which separates the (intracapsular) lymph space from the surrounding interstitial space. The spindle is richly supplied with afferent and efferent nerve fibres which are of six types (two afferent, four efferent; see below) comprising up to 20 individual fibres. The afferent neurones originate in the central spindle region from sensory endings located on the intrafusal fibres, well inside the capsule. This central, or sensory, region is conventionally subdivided into the equatorial region (occupied by the primary endings of the large and fast-conducting Ia afferents) and the juxtaequatorial region (occupied by the secondary endings of the smaller and more slowly conducting group II afferents; see Fig. 3) (*Banks* et al. 1982; *Barker* 1974a; *Boyd* 1981a; *Matthews* 1981a,b,c; *Kennedy* et al. 1980; *Taylor* and *Prochazka* 1981).

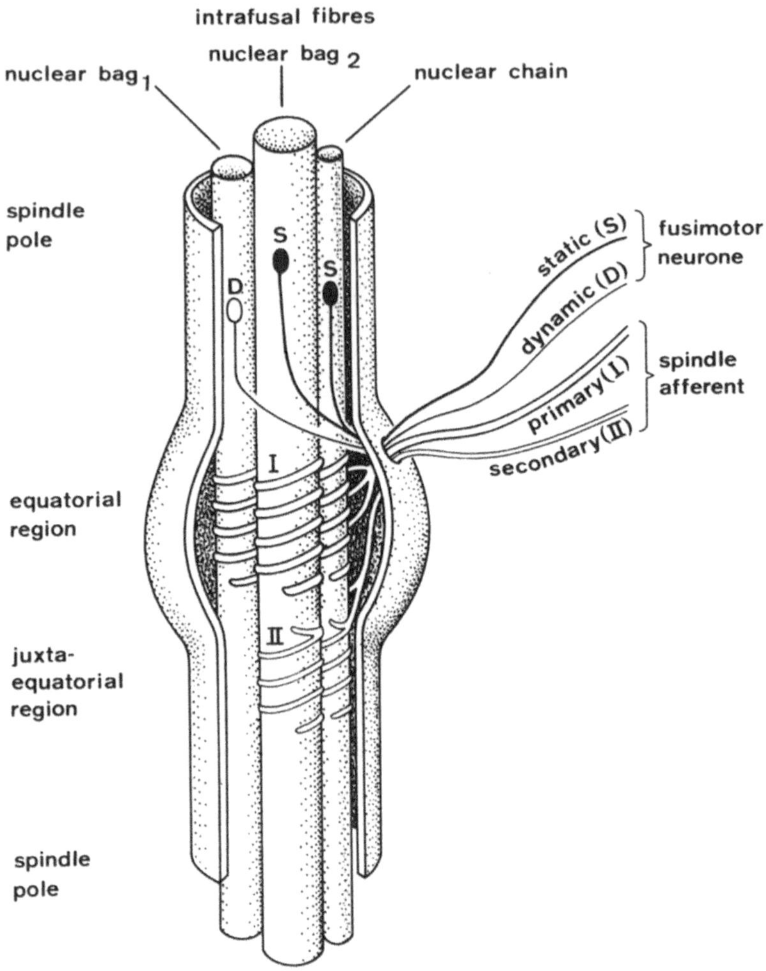

Fig. 3. Simplified scheme of the current picture of the mammalian spindle. Each type of intrafusal muscle fibre shown by a single representative, although especially nuclear chain fibres occur in much larger numbers. The static and dynamic fusimotor neurones may be either exclusively fusimotor (γ-fibres) or mixed skeletofusimotor (β-fibres). The static innervation pattern illustrated (to bag_2 and chain fibres) applies for γ_S-fibres, but not generally for β_S-fibres, since the latter are more selective by supplying the longest of the chain fibres (not illustrated). The motor terminals are merely depicted as functional connections, without intentional resemblance to any of the current schemes (see text, Sect. 2.2.1.4). (Modified from *Schmidt* 1978, with permission)

2.2.1.2 Intrafusal Muscle Fibres. The intrafusal muscle fibres are of three types: *bag_1*, *bag_2*, and *chain fibre* (Fig. 3). These may be distinguished by their light- or electron-microscopic characteristics (Table 1) and by their mechanical and electrical properties (Table 2). The bag_2 fibre occupies an intermediate position between the bag_1 and the chain fibres. Morphologically it is somewhat closer to the bag_1 fibre, yet in its histochemical

Table 1. Structure and innervation of intrafusal fibres (*Boyd* 1981a, with permission)

	Dynamic bag_1 fibre	Static bag_2 fibre	Chain fibres
Number	1 (occasionally 2)	1	3–5
Equatorial nuclei	Nuclear bag	Nuclear bag	Nuclear chain
Length	Intermediate (8 mm)	Long (9 mm)	Short (4 mm)
Diameter	Large (20 μm)	Large (25 μm)	Small (12 μm)
Rate of atrophy when denervated	Slow	Slow	Fast
ATPase activity (alkaline pre-incubation)	Low	Medium/high	High
Glycogen content	Low/medium	Medium	High
Elastic fibres at poles	Scarce	Prominent	Present
Electron microscopy distinct single M-line	Absent (except extracapsular region)	Present (except equatorial region)	Present
Development	Second fibre formed	First fibre formed	Last fibres formed
Motor innervation	Dynamic γ- ($\pm$ β-) axons	Static γ-axons	Static γ- ($\pm$ β-) axons

Table 2. Mechanical and electrical properties of intrafusal fibres (*Boyd* 1981a, with permission)

	Dynamic bag_1 fibre	Static bag_2 fibre	Chain fibres
Contraction when axon stimulated at 10 Hz	Smooth	Small, smooth	Oscillatory
Stimulation frequency for maximal contraction	75–100/s	100/s	150–200/s
Tetanic contraction time	1.0 s	0.6 s	0.4 s
Amplitude of movement	Small	Large	Large
Maximal extension of sensory spirals	2%–8%	12%–30%	15%–20%
Creep back in spirals following stretch of active fibre	20%–30%	Usually absent	Absent
Electrical response at motor nerve endings	Local response, non-propagated	Local response, non-propagated	Propagated action potential
Response to topically applied acetylcholine	Contraction (most sensitive)	Contraction (less sensitive)	No contraction (neuromuscular block)

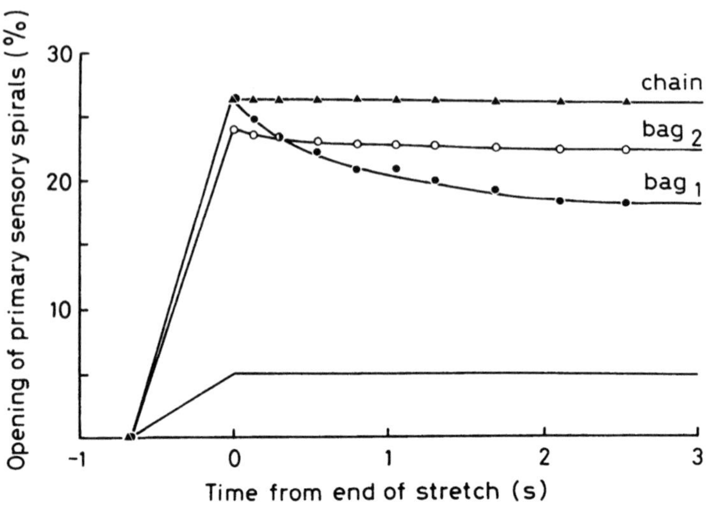

Fig. 4. Time course of extension of the sensory region, as measured by the degree of opening of the sensory spirals of a primary ending when the spindle was exposed to a ramp-and-hold stretch (*bottom trace*). Presence of creep in the bag_1 fibre (*filled circles*) but not in the bag_2 fibre (*open circles*) nor in the chain fibres (*triangles*). Isolated spindle preparation. (From *Boyd* 1976b, with permission)

(Table 1) and mechanical (Table 2) properties it is much closer to the chain fibres (*Banks* et al. 1977b, 1982; *Barker* 1974a; *Barker* et al. 1978; *Bessou* and *Pagès* 1972; *Boyd* 1976a,b, 1981a; *Cooper* and *Gladden* 1974; *Gladden* 1976; *Kennedy* et al. 1980; *Kucera* 1981a, 1982a,f; *Smith* and *Ovalle* 1972).

For some of the characteristics used for the differentiation of the various fibre types, their very property of being convenient markers for identification may be their sole function. For the majority of these markers it is still difficult to assess just how important, quantitatively, they are to account for the functional properties of the different types of intrafusal fibres. For instance, the particular arrangement of the fibre nuclei in the equatorial region may mean no more than that bag fibres are of larger diameter than chain fibres. Yet for all types of fibre the accumulation of nuclei goes hand in hand with a reduced density of contractile material (myofibrils). This is bound to affect the stiffness (most likely by reducing it) of the sensory region. Also, it remains to be seen to what extent the gradation of ultrastructural properties along the bag_2 fibre can account for the specific role of this fibre as the carrier of dynamic fusimotor action (see Sect. 2.2.1.4; also *Matthews* 1981a). It bears emphasis that longitudinal gradation of properties, as such, is common to all three types of intrafusal fibres (*Banks* et al. 1977a). This holds true especially for the activity of myofibrillar ATPase (*Barker* et al. 1976a; *Banks* et al. 1977a; *Kucera* 1977, 1981b).

On the other hand, it is hard to believe that the histochemical characteristics, which so clearly discriminate among the three types of fibre (Table 1), should be irrevant for their contraction properties (Table 2). Especially the activity of the myofibrillar ATPase should be influential in determining the speeds of contraction of the different fibre types (see *Barany* 1967). However, the histochemical tests currently employed tell us more about the enzyme's capability to resist alkaline pre-incubation than about its normal activity at a physiological pH. This may explain why the large differences in

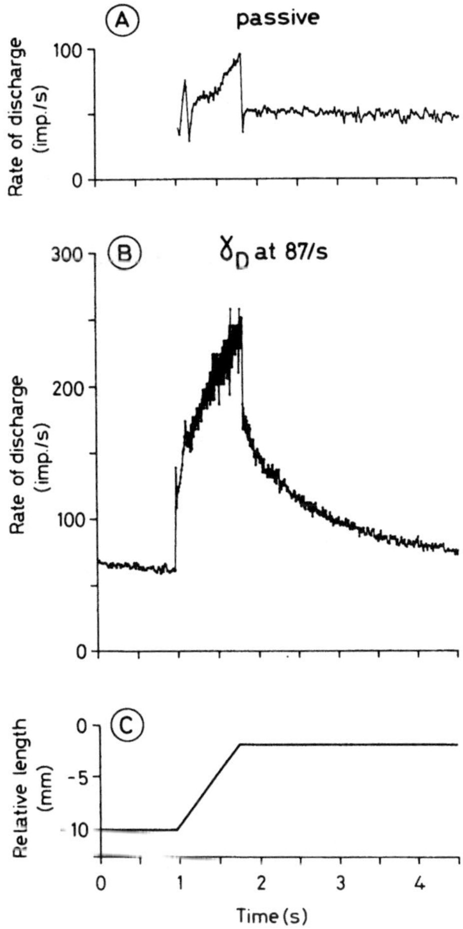

Fig. 5A–C. Responses of a primary spindle afferent from the soleus muscle to a ramp-and-hold stretch applied to the parent muscle (**C**). In **A**, passive response with initial burst and deceleration response at the beginning and end of the ramp, respectively. In **B**, pronounced enhancement of the dynamic response during the ramp phase when a single dynamic γ-axon was stimulated tonically (at 87/s). Note slow decay (creep) during the first 2 s after completion of the stretch. (From *Hulliger* 1976)

alkaline ATPase activity are not accompanied by large differences in contraction speed: the tetanic contraction speeds of chain and bag_1 fibres differ only by a factor of about two (*Boyd* 1976a,b; *Boyd* et al. 1977; Table 2) and in the responses of primary afferents such differences are hardly detectable (see *Hulliger* 1979 and Sect. 3.2.3). Such reservations were borne out by the recent immunohistochemical demonstration (in the rat) of the existence of fast myosin in chain fibres and also in bag_1 but not in bag_2 fibres, whereas slow myosin was absent in chains but present in both types of bag fibre, in the bag_2 fibre even to the exclusion of the fast enzyme (*Celio* et al. 1981). On this evidence, the bag_2 fibres could hardly be a faster contractile system than the bag_1 fibres.

Among the mechanical properties (Table 2), two seem functionally important. First, the amplitudes of movement at the contraction site as well as the size of the extension of the sensory region ('spirals' of the afferent endings, in Table 2) differ considerably between bag_1 fibres (small) and bag_2 and chain fibres (large). This agrees with the smaller excitatory action of dynamic axons (operating bag_1 fibres; see below) when compared with static axons (operating bag_2 and chain fibres). Second, the response of primary spindle afferents to a ramp-and-hold stretch may show a slow adaptive decay (creep; Table 2), lasting 1–2 s, after the pronounced dynamic response during the phasic stretch (see Fig. 8A). This adaptive creep is very much augmented during tonic activation of dynamic fusimotor fibres (Figs. 5 and 8C). *Boyd* (1976b;

Boyd et al. 1977) has shown that this is a distinguishing feature of the bag_1 fibre, both in the passive state (see Fig. 4) and during dynamic activation (*Boyd* et al. 1977).

This creep could reflect the gradual decay of an active state of the bag_1 fibre, which may have been induced or facilitated by the preceding dynamic stretching of the muscle, not unlike the stretch activation described for insect fibrillar flight muscles (*Pringle* 1949; see also *Pringle* 1978). Recent observations on the passive bag_1 fibre have indeed provided evidence for the occurrence of such stretch activation (*Poppele* and *Quick* 1981), and this process may also be responsible for the sensitizing action of dynamic fusimotor fibres, as has repeatedly been suggested (*Boyd* et al. 1977; *Hulliger* et al. 1977a; also *Poppele* and *Quick* 1981).

2.2.1.3 Afferent Neurones. The spindle is innervated by two types of sensory neurones: the Ia or primary afferent (one, sometimes two per spindle) and the group II or secondary afferent (zero to five per spindle). These branch widely and supply the primary endings in the midequatorial region and the secondary endings in the juxtaequatorial regions on either side, where the latter may intermingle with motor terminals. The primary endings regularly occur on all three types of intrafusal fibre (Fig. 3), and accordingly, the Ia afferents are normally excited by both static and dynamic fusimotor neurones (see below). The group II afferents are preferentially situated on bag_2 and chain fibres and normally operated by static fusimotor neurones only (*Banks* 1981; *Banks* et al. 1977a, 1979, 1981a, 1982; *Bessou* and *Pagès* 1969; *Kennedy* et al. 1980; *Kucera* 1982d,e; *Matthews* 1972, 1981a,b,c; *Quick* et al. 1979).

In an early study (*Appelberg* et al. 1966) dynamic γ-axons, identified by their action on the primary afferent of the same spindle, were found to only rarely act on secondary afferents. One should therefore expect to find secondary endings on nuclear chain and bag_2 fibres only (see Fig. 3). However, this has not been borne out by recent anatomical investigations using serial reconstruction and tracing techniques (*Banks* et al. 1979, 1981a, 1982). Depending on the muscle studied, up to 73% of the bag_1 fibres were found to receive terminal branches from secondary endings. Yet the influence from bag_1 – and dynamic fusimotor – fibres on secondary afferents may be less pronounced than this figure suggests, since the bag_1 contact area of secondary endings probably is less than 10% of the total contact area on intrafusal fibres (*Banks* et al. 1981a, 1982). Functionally, the selectivity of static action on secondary afferents may, therefore, be relative rather than absolute.

For most muscles an axonal conduction velocity of 72 m/s appears to be a useful dividing line to differentiate between primary and secondary afferents. However, this is not an absolute criterion, since for cat masseter muscle the equivalent figure probably is around 45 m/s (*Inoue* et al. 1981; *Morimoto* et al. 1982; see also *Appenteng* et al. 1980).

2.2.1.4 Motor Innervation. The motor innervation of the spindle is shared between γ- and β-motoneurones. γ-Axons are slowly conducting (below 55 or 50 m/s) and their sole known function is fusimotion. In contrast, β-axons have conduction velocities in the α-range (above 50 m/s) and they are mixed skeleto- and fusimotor (but never exclusively fusimotor), since they supply both extrafusal and intrafusal muscle fibres. Both *β- and γ-motoneurones* can functionally be further subdivided into *static* and *dynamic types,* according to whether they decrease or increase the dynamic response of Ia afferents to large stretches. Whilst static and dynamic β-motoneurones can also be differentiated on the basis of their axonal *conduction velocity* (static motoneurones above and dynamic motoneurones below 85 m/s), no similarly convenient criterion has so far been established to differentiate between the two types of γ-motoneurones. In particular, whenever the matter was investigated, the axonal conduction velocities of static and dynamic γ-fibres were found to overlap.

Dynamic fusimotor neurones (both β and γ) innervate nuclear *bag*$_1$ fibres selectively. *Static* γ-motoneurones supply either *bag*$_2$ or chain fibres or both, whereas static β-motoneurones seem to be more selective in that they preferentially innervate the longest chain fibre. *Dynamic* γ-motoneurones act selectively on primary afferents. In contrast, *static* γ-motoneurones operate *primary and secondary* afferents indiscriminately. This is explicable from present knowledge of the motor innervation of the intrafusal fibres and of the location of the sensory terminals on these fibres (see Sect. 2.2.1.3 and Fig. 3). For β-axons a similar distinction concerning their action on secondary afferents seems not to be as clear-cut as for γ-axons.

In a recent study about a quarter of those secondary afferents from peroneus tertius which were subjected to β-action were operated by dynamic β-axons, as classified by their action on primary afferents of the same muscle (*Jami* et al. 1982). The dynamic β-effects on the secondary afferents were, though, appreciably weaker than the static β-effects. Yet even *static* β-action on either type of afferent fibre may turn out to be weaker than *static* γ-action, owing to the restriction of static β-axons to a single chain fibre.

The motor organization of the muscle spindle is further characterized by both *convergence and divergence.* Individual fusimotor neurones may act on up to seven different spindles in the same muscle and individual spindles may be operated by up to ten fusimotor neurones, with two dynamic and three to five static fibres being the rule. Since even single fusimotor neurones can exert powerful effects on the discharge rate and sensitivity of spindle afferents, the fusimotor system, if activated at its full physiological capacity, must be a very powerful tool for the efferent control of the sensory feedback from muscle spindles (*Andrew* et al. 1978; *Bessou* et al. 1966; *Boyd* et al. 1977; *Boyd* 1981a; *Brown* et al. 1965; *Brown* and *Matthews* 1966; *Crowe* and *Matthews* 1964b; *Ellaway* et al. 1972; *Emonet-Dénand* et al. 1977b, 1980a; *Hunt* 1974; *Hunt* and *Kuffler* 1951a; *Jami* et al. 1982; *Kennedy* et al. 1980; *Kucera* and *Hughes* 1983; *Kuffler* and *Hunt* 1952; *Matthews* 1962, 1972, 1981a,b; *Petit* et al. 1983).

Motor Terminals. Functionally the morphology of the terminals of fusimotor neurones may be largely irrelevant, since even the generation of either propagated or local potentials in intrafusal fibres (*Bessou* and *Pagès* 1972) may be determined by the properties of the extra- rather than the subjunctional membrane. Nevertheless, if shape and ultrastructure of the terminals were solely dependent on the type of motor axon, these might serve as useful markers in studies of the motor supply to the various subtypes of intrafusal fibres (cf. *Boyd* 1962; *Barker* et al. 1970). However, the weight of recent evidence no longer supports this notion. *Kucera* (1982a,b), *Kucera* and *Hughes* (1983) and *Arbuthnott* et al. (1982) have emphasized the importance of intrafusal fibres and even of sensory endings in determining shape and ultrastructure of motor terminals, in addition to any influence from the motor axon. This is in keeping with the relative unimportance of motor innervation for the normal development of intrafusal muscle fibres (*Zelená* and *Soukup* 1973, 1974; *Kucera* 1981b, 1982g; but cf. *Te Kronnie* et al. 1982).

Apart from this the issue of fusimotor terminal morphology is confused. Three schemes obtained with different techniques are currently debated, yet these schemes do not match up. It remains to be fully established to what degree the histochemical categories of cholinesterase rims and plates (*Kucera* 1980a,b,c, 1981b, 1982a,b) conform with the four ultrastructural types of *Arbuthnott* et al. (1982). More seriously, both these schemes are at least partly incongruent with the classification of motor terminals as p_1, p_2 or trail endings (*Barker* et al. 1970, 1976a, 1978; *Banks* 1981). Especially the concept of p_1 plates as uniform markers of both static and dynamic β-innervation has been challenged as untenable by *Arbuthnott* et al. (1982) and by *Kucera* and *Hughes* (1983). It seems that only the use of combined histological, ultrastructural and morphometric techniques, as well as mutual agreement on terminology, will help clarify this and several other discrepancies.

Autonomic Innervation. The innervation of intrafusal muscle fibres by autonomic axons, in addition to the established vasomotor supply to the spindle, has long been equivocal (see *Matthews* 1972; *Barker* 1974a; *Hunt* 1974). However, recent fluorescence and electron-microscopic evidence has now firmly established an autonomic motor supply to intrafusal muscle fibres (*Ballard* 1978; *Banks* 1981; *Barker* and *Saito* 1981). Both noradrenergic and cholinergic terminals appear to be involved, probably acting on all types of intrafusal fibre indiscriminately. Moreover, the autonomic innervation seems non-selective, since it is mostly derived from vasomotor axons (*Barker* and *Saito* 1981).

A functional study has since corroborated these findings (*Hunt* et al. 1982) when weak biasing effects (in the absence of fusimotor stimulation) and weak facilitation of fusimotor action on spindle afferents were found upon stimulation of sympathetic fibres. These were short-latency effects, which were, however, offset by sufficiently long-lasting vasoconstriction causing spindle ischaemia. On this evidence, the functional significance of the autonomic spindle innervation remains difficult to assess. The findings of long-latency excitation upon sympathetic stimulation (*Passatore* and *Filippi* 1981, 1982; *Filippi* et al. 1983) seem to bear on this issue, yet it cannot be excluded that these observations were due to reflex effects on the (intact) fusimotor supply (see *Hunt* et al. 1982).

2.2.2 Pending Issues and Growing Points

The distribution of muscle spindles throughout their parent muscle is often non-uniform (see *Matthews* 1972; *Barker* 1974a). In addition, muscle spindles may occur in close association with other muscle receptors, forming either *tandem* spindles (*Barker* and *Ip* 1961; *Banks* et al. 1982; *Bakker* and *Richmond* 1982), *dyads* with Golgi tendon organs (*Richmond* and *Abrahams* 1975, 1979b; *Lund* et al. 1978; *Richmond* and *Bakker* 1982), or spindle *complexes* containing up to ten individual spindles (*Richmond* and *Abrahams* 1979b; *Bakker* and *Richmond* 1982).

The function of such arrangments remains a mystery. Perhaps isolated spindle studies will reveal specific functional properties, or else tracing methods might demonstrate specific central projections. Failing that, such configurations may turn out to be mere developmental epiphenomena.

Nuclear chain fibres are no longer considered as a uniform group of muscle fibres. *Long chain fibres* have been singled out on morphological grounds (*Barker* et al. 1976a; *Kucera* 1980b,c, 1982c) and it has become apparent that histochemically they differ from the majority of (short) chain fibres (NADH-TR reaction, but not ATPase reaction; *Kucera* 1980b). *Kucera* (1980b, 1982a,c,f) further distinguishes (morphometrically) a group of intermediate chain fibres. Whilst long chain fibres seem preferentially innervated by static β-axons (see Sect. 2.1.5), nothing is known about any specific innervation of the intermediate fibres. Yet if the morphological variety of the muscle fibres supplied by static axons (bag_2, long, intermediate and short chains) were paralleled with a comparable range of mechanical (or transducer) properties, this might account for the range of static fusimotor effects described below (Sect. 3.2.2).

The investigation of intrafusal fibre properties might profit from the use of special models. In cat hindlimb muscles the majority of spindles are equipped with one bag_1 and one bag_2 fibre each. Yet between 5% and 10% of the spindles (mostly in tandem configuration) lack a bag_1 fibre (*Kucera* 1981a, 1982a,f; *Banks* et al. 1982). In neck muscles the fraction of such *bag_2 spindles* is much higher (*Bakker* and *Richmond* 1981, 1982; *Abrahams* 1981). Interestingly, in these muscles a higher-than-normal proportion of afferents with intermediate dynamic response characteristics has also been found (*Richmond* and *Abrahams* 1979a). These muscles might therefore provide useful preparations for combined anatomofunctional studies on the contribution of bag_2 and chain fibres to different components of the Ia response.

A nearly complementary model is provided by certain spindles which during the early stages of regeneration after nerve injury fail to give maintained responses to static stretch of the parent muscle (*Brown* and *Butler* 1976; *Hyde* and *Scott* 1983). Instead they only respond with transient afferent discharge during the dynamic phase of a stretch. Such spindles might provide a tool for the assessment of the structural and functional properties of the *bag_1 fibre* and its contribution to the dynamic sensitivity of primary afferents.

The future of research on the structure of the muscle spindle probably lies in the continued pursuit of an interdisciplinary approach. With the issue of spindle innervation largely settled in its essentials, such combined efforts could now be directed at some of the more general questions of receptor physiology. These include the processes responsible for the pronounced dynamic sensitivity of the primary afferent, as well as the mechanisms underlying the control of this sensitivity by fusimotor neurones.

2.3 Morphology of the Human Spindle: Comparative Aspects

In several respects, data on spindle behaviour during voluntary movement in man are at variance with those obtained in cat and monkey. This might be due to structural idiosyncrasies of human spindles. Yet this is not so far borne out by morphological evidence.

2.3.1 Structural Similarities

The main conclusion from morphological studies on human spindles is that they are structurally very similar to those of the cat (*Cooper* and *Daniel* 1963; *Kennedy* 1970; *Rakhawy* et al. 1971; *Swash* and *Fox* 1972; *Barker* 1974a; *Harriman* et al. 1974; *Kubota* and *Masegi* 1977; *Kucera* and *Dorovini-Zis* 1979). Thus, human spindles appear to have a normal sensory innervation and they usually are generously supplied with intrafusal muscle fibres (up to 14; *Cooper* and *Daniel* 1963; *Swash* and *Fox* 1972), which comprise bag_1, bag_2 and chain fibres (*Harriman* et al. 1974; *Kucera* and *Dorovini-Zis* 1979). Furthermore, a rich motor supply to intrafusal muscle fibres has been described (for details, see *Prochazka* and *Hulliger* 1983). However, the detailed fusimotor innervation pattern as it has emerged from the cat studies (see Sect. 2.2.1.4 and *Boyd* 1981a; *Matthews* 1981a) remains to be confirmed in man.

As regards the incidence of *β-innervation in man,* the morphological evidence available indicates that it might be as widespread as in the cat, since large myelinated axons have regularly been found to supply p_1 plates on bag fibres (*Swash* and *Fox* 1972; *Barker* 1974a; *Saito* et al. 1977; cf. also *Cooper* and *Daniel* 1963). There is, however, some uncertainty, since like in the cat it is not clear whether *all* p_1 plates are innervated by β-axons or whether some might also be supplied by γ-axons (*Barker* et al. 1980; cf. also Sect. 2.2.1.4).

2.3.2 Structural Differences

Structural differences between human and feline spindles have also been encountered. Thus, *Swash* and *Fox* (1972) but not *Cooper* and *Daniel* (1963) found primary endings mainly on nuclear bag fibres. This clearly is at variance with the sensory innervation described for the cat (Sect. 2.2.1.3), where primary terminals are abundant also on chain fibres (*Barker* 1974a; *Banks* 1981; *Banks* et al. 1979, 1981a, 1982). The human studies referred to were performed before the existence of the three types of intrafusal fibre was generally accepted (*Barker* et al. 1976a; *Boyd* 1976a,b). Thus it is conceivable that in man primary afferents might be more responsive to activation of bag_1 than of bag_2 and chain fibres and, therefore, perhaps more susceptible to dynamic than to static fusimotor activity. Further, human spindles contain more intrafusal muscle fibres than were encountered in other mammals (*Cooper* and *Daniel* 1963; *Barker* 1974a). This is largely accounted for by larger numbers of nuclear chain fibres (*Cooper* and *Daniel* 1963; *Kucera* and *Dorovini-Zis* 1979). Static fusimotor action on secondary afferents might, therefore, be more powerful in man than in the cat.

In man, a closer topographic relationship than in the cat may exist between spindles and extrafusal motor units. The capsule of human spindles can enclose neighbouring extrafusal muscle fibres (*Cooper* and *Daniel* 1963; *Kucera* and *Dorovini-Zis* 1979). Thus human spindle afferents might be even more responsive to the contraction of individual motor units than spindle afferents in the cat, where such responsiveness to twitches in single motor units has repeatedly been emphasized (*Binder* et al. 1976; *Windhorst* and *Meyer-Lohmann* 1977; *Bottermann* et al. 1978; *Binder* and *Stuart* 1980; *Cameron* et al. 1981). These findings have even led to the suggestion that generally muscle spindle afferents are perhaps more concerned with the monitoring of mechanical events in small muscle compartments than with the recording of overall muscle length. A recent study in man (*McKeon* and *Burke* 1983) has indeed confirmed such responsiveness to single motor unit contractions, but a systematic quantitative comparison in terms of sensitivity to force modulation has not so far been possible.

3 Functional Properties of Spindle Afferent and Fusimotor Neurones

Reduced to the bare essentials, the distinguishing functional properties of the muscle spindle neurones are: first, that the primary afferent possesses high dynamic sensitivity for muscle stretching in addition to its static (or position) sensitivity; second, that for the secondary afferent static sensitivity is more pronounced than dynamic sensitivity; third, that dynamic fusimotor neurones further increase the dynamic sensitivity of primary afferents; and fourth, that static fusimotor neurones, in addition to their powerful excitatory effect, reduce dynamic sensitivity more clearly for primary than for secondary afferents. Yet, these are crude simplifications asking for considerable refinement, especially in the light of recent findings.

3.1 Stretch Receptor Properties of the Muscle Spindle

3.1.1 Dynamic Sensitivity: Primary, Secondary and Intermediate Afferents

Muscle spindle afferents are readily excited by stretches applied to the parent muscle. When the stretch is maintained at a constant muscle length the responses of primary and secondary afferents are much the same, apart from the higher variability of discharge which is typical of primaries (*Matthews* and *Stein* 1969b). Thus the two types of afferents do not drastically differ in their *position sensitivity* (see *Matthews* 1972), although the values found for secondaries tend to be higher (by a factor of 1.5) (*Lennerstrand* 1968a; *Appelberg* et al. 1982a; *Bottermann* and *Eldred* 1982).

Functionally, the main distinguishing feature between primary and secondary afferents lies in the difference in *dynamic sensitivity* to muscle stretch. Thus, during the dynamic phase of a ramp-and-hold stretch, primary afferents give a transient (i.e. velocity) response which is much more pronounced than that of secondary afferents (*Cooper* 1961; *Harvey* and *Matthews* 1961; *Matthews* 1963, 1972/references; also *Houk* et al. 1981). In general terms, the responses of primary afferents reveal sensitivity to position, velocity and acceleration, whereas secondary afferents respond only to the position and velocity components of the stimulus (*Matthews* and *Stein* 1969a; *Poppele* and *Bowman* 1970; *Goodwin* et al. 1975; *Cussons* et al. 1977). The acceleration sensitivity of the primary afferent has been unambiguously demonstrated in studies on the frequency response, as measured with small sinusoidal stretches (references above). Yet it is also manifest in responses to large ramp-and-hold stretches, as a clear prolongation of the first, compared with subsequent, interspike intervals at the end of the dynamic phase of the stretch (deceleration response, *Matthews* 1972; *Cheney* and *Preston* 1976a; also Figs. 5A, 8A,B). On the other hand, the initial burst of discharge at the beginning of a ramp stretch is not a reliable indicator of acceleration sensitivity (see *Brown* et al. 1969a; *Prochazka* and *Hulliger* 1983).

There is considerable scatter in the relationship between axonal conduction velocity and dynamic response (*Matthews* 1963; *Koeze* 1968, 1973a; *Cheney* and *Preston* 1976a), which is partly due to the existence of spindle afferents with intermediate response properties (*Rack* and *Westbury* 1966; *Browne* 1975; *Dutia* 1980; *Inoue* et al. 1981). Often these *intermediate spindle afferents* also have intermediate values of axonal conduction velocity (around 72 m/s; but see Sect 2.2.1.3). However, in neck muscles units with intermediate dynamic properties and conduction velocities well above 72 m/s are not exceptional (*Richmond* and *Abrahams* 1979a; *Abrahams* 1981). Differences in size of locally effective stretch

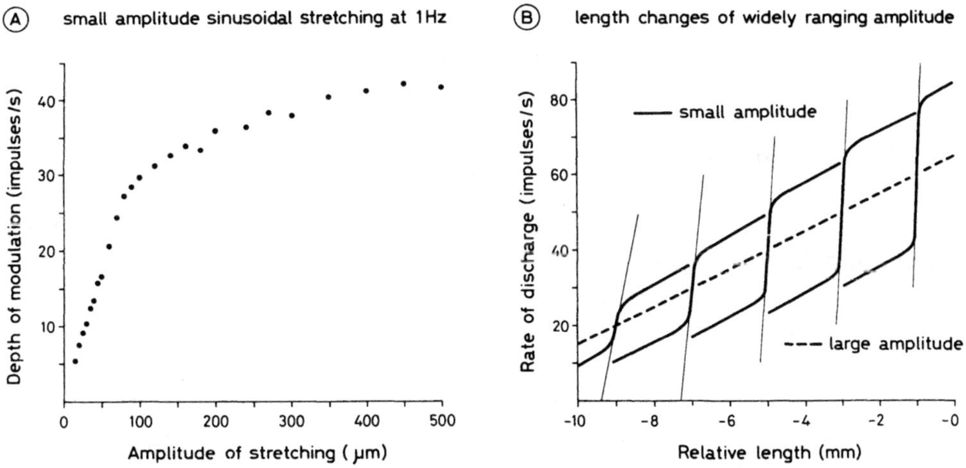

Fig. 6A,B. Small-movement sensitivity and static-position sensitivity of primary spindle afferents. In **A**, the magnitude of a primary unit's responses to sinusoidal stretching is plotted as depth of modulation against the half peak-to-peak amplitude of stretching. In the linear range the sensitivity (or gain), expressed as modulation/amplitude of stretch, was 450 impulses/s/mm. **B** Schematic illustration of the relationship between static-position sensitivity (as measured from responses to large and maintained stretch [*broken line*]) and small movement sensitivity (from responses to sinusoidal stretching at different mean muscle lengths [*fine lines*]). Position sensitivity: 5 impulses/s/mm; small-movement sensitivity: 50–250 impulses/s/mm, incrementing in steps of 50 (figures in broad agreement with experimental findings). (From *Hulliger* 1981, with permission)

may further account for some of the scatter at issue (*Meyer-Lohmann* et al. 1974; *Windhorst* et al. 1975, 1976; *Jami* and *Petit* 1979; *Banks* et al. 1981b).

3.1.2 Small-Movement Sensitivity

The hope behind the first application of systems analysis methods was that the spindle might lend itself to general description by linear transfer functions (*Matthews* and *Stein* 1969a; *Poppele* and *Bowman* 1970). However, this expectation has not been fulfilled, since spindle responses revealed remarkable non-linearities (see Sect. 3.1.3). Nevertheless, these studies have led to important insights, e.g. the recognition of the primary afferents' extraordinary sensitivity to movements of very small amplitude (*Matthews* and *Stein* 1969a).

For a narrow range of stretch amplitudes, the so-called *linear range* (up to 100 μm for sinusoids, at 1 Hz), the sensitivity (i.e. the response/stimulus ratio) of a primary afferent tends to be approximately constant, when response increases proportionally with the amplitude of stretch (see Fig. 6A). Yet outside this linear range sensitivity declines progressively as the size of the stimulus is further increased. This is referred to as gain-compression non-linearity (see *Hunt* and *Wilkinson* 1980). Such non-linear behaviour is typical for both primary and secondary afferents, either when they are passive or during activation of static or dynamic fusimotor fibres (*Matthews* and *Stein* 1969a; *Chen* and *Poppele* 1973; *Hasan* and *Houk* 1975a,b; *Cussons* et al. 1977; *Hulliger* et al. 1977a,b; cf. also Fig. 10).

For primary afferents the values of sensitivity in the small-movement linear range are of the order of 500 impulses/s/mm (*Goodwin* et al. 1975; *Hasan* and *Houk* 1975a; *Baumann* and *Hulliger*, to be published a; cf. also Fig. 7B). Thus, small-movement sensitivity may be up to 100 times higher than static-position sensitivity. Passive secondary afferents are about five times less sensitive than primaries (100 impulses/s/mm), yet static fusimotor action largely abolishes this difference (*Cussons* et al. 1977). Thus, spindle afferents – and particularly primary units – are highly sensitive detectors of small movements, whilst their responses to large stretches are comparatively poor.

The discrepancy between small- and large-movement sensitivity is shown schematically in Fig. 6B. The slope of the broken line indicates position sensitivity, and the steeper slopes of the fine, straight lines reflect the higher values of small-movement sensitivity, which tend to increase with mean muscle length (see also Sect. 3.1.3). Thus, the diagram illustrates the finding that the spindle is capable of resetting its small-movement sensitivity to high levels over the entire physiological range of muscle lengths.

Such resetting has been attributed to the formation and persistence, even in relaxed muscle, of stuck cross-bridges between intrafusal actin and myosin filaments, and the decline of sensitivity with increasing amplitude of movement is thought to be due to a progressive breakdown of these cross-bridges (*Brown* 1971b; *Matthews* 1972; cf. *Hill* 1968; *Brown* et al. 1969a). Thus, the high sensitivity to small movements and the phenomenon of gain compression are largely regarded as properties of the mechanical filter (see Sect. 2.1.4). This view has been strongly supported by direct measurements of mechanical events (vis. intrafusal tension) in isolated spindle preparations (*Hunt* and *Wilkinson* 1980; cf. also *Hunt* and *Ottoson* 1976).

High values of small-movement sensitivity have regularly been found in studies using minute stretches which were applied at fixed muscle length (references above). However, this bears little resemblance to natural movements, which mostly fall well outside the narrow linear range. In view of the gain compression behaviour of the spindle, it was therefore an open question whether a high sensitivity for small disturbances would be preserved during movements of larger amplitude. Recent data obtained with combined (small sinusoidal and larger triangular) stretches have shown that under such conditions small-movement sensitivity could be drastically

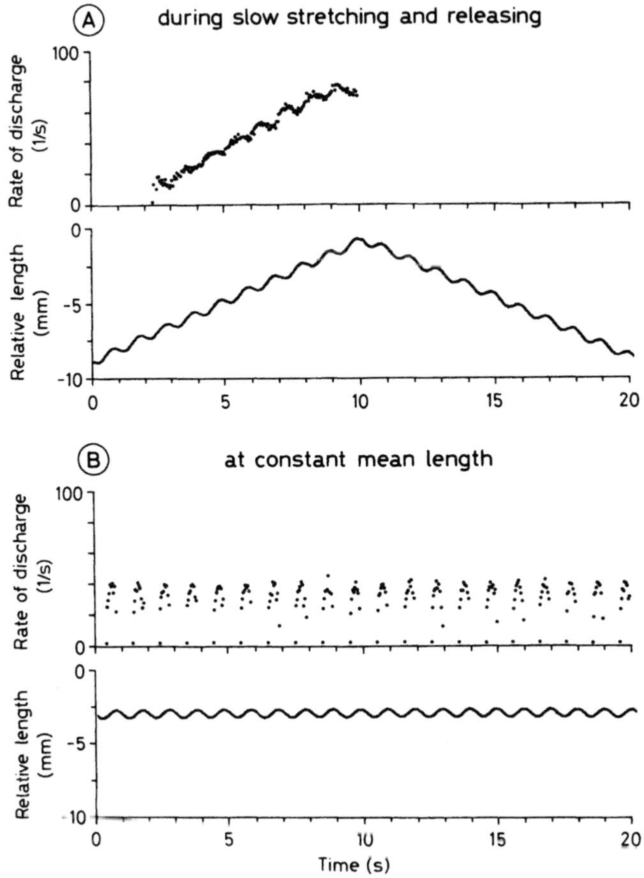

Fig. 7A,B. The response of a Ia afferent unit from cat soleus muscle to sinusoidal stretching at 50 μm. **A** Sinusoids superimposed on large triangular movement with a peak-to-peak excursion of 8 mm (reaching up to 1 mm below maximum physiological length). **B** Sinusoidal stretching at constant mean muscle length (3 mm below the physiological maximum). *Lower tracings* (in **A** and **B**) are schematic illustrations of the time course of muscle stretch with a disproportionally large display of the sinusoidal component (increased fivefold for clarity). (From *Hulliger* 1981, with permission)

reduced (to between 5% and 20% of the control values at constant muscle length; *Baumann* and *Hulliger*, to be published a). In the example of Fig. 7, sensitivity to sinusoidal stretch during concomitant triangular movement (Fig. 7A) did not exceed 75 impulses/s/mm, as compared with 780 impulses/s/mm at constant muscle length (Fig. 7B). Moreover, it was found that the extent of this reduction in sensitivity during movement depended on the velocity of the movement and that the high values of sensitivity which are encountered at constant mean length were only manifest during very slow movements (below 0.1% resting length/s).

These findings provide another example of non-linear spindle behaviour, since for a linear system the response to the sum of two stimuli should be the sum of the responses to the individual stimuli (superposition principle), and the sensitivity to small sinusoids should be independent of the velocity of larger superimposed movements. Thus, in this case primary afferents violate the superposition principle. For secondary afferents as well as for primaries and secondaries exposed to fusimotor activity, this matter remains to be investigated.

These findings also indicate that the putative role of spindle primary afferents as particularly sensitive detectors of small disturbances during voluntary movement (see *Matthews* 1972; *Stein* 1974) may be more questionable than previously thought: unless a muscle is moved very slowly, the small-movement sensitivity of passive primary afferents is no higher than for movements well outside the classical linear range (*Baumann* and *Hulliger,* to be published a). Thus, under passive conditions high-sensitivity monitoring of small disturbances would be restricted to postural tasks.

3.1.3 Non-linear Properties of the Muscle Spindle

Non-linear spindle behaviour is of more than formal analytical interest. It complicates the use of mathematical modelling to evaluate the role of spindle afferents in the control of movement (see, e.g. *Poppele* 1981; *Hasan* 1983). At a more qualitative level of analysis it narrows the scope of deductions from spindle afferent discharge to arrive at estimates of fusimotor activity during natural movements (see Sect. 5.2). Examples of non-linear spindle behaviour, as seen in passive afferents or during fusimotor activation, are listed below.

Violation of Superposition Principle. For stimuli in the small-movement linear range the superposition principle applies for spindle afferents, since frequency response data, as obtained with sinusoids, can be used to correctly predict the responses to other stimuli such as rectangular or triangular stretches (*Poppele* and *Bowman* 1970; *Hasan* and *Houk* 1975a). However, for stretches outside the linear range this does not hold true any longer, since the response to a given sinusoid then depends on the size and/or frequency content of additional stimulus components (*Hasan* and *Houk* 1975a; *Chen* and *Poppele* 1978; *Poppele* 1981; *Matthews* and *Watson* 1981; *Baumann* and *Hulliger,* to be published a; see also Sect. 3.1.2).

Gain Compression. The gain-compression non-linearity detailed above (Sect. 3.1.2) has been encountered in numerous sensory systems (see *Thorson* and *Biederman-Thorson* 1974). For the spindle it is bound to contribute – in combination with other likely mechanisms such as pacemaker switching (Sect. 2.1.4) – to violations of the superposition principle (*Poppele* 1981).

Length Dependence. The sensitivity to small movements, in addition to being orders of magnitude higher than static-position sensitivity (Sect. 3.1.2), also depends on mean muscle length (*Matthews* and *Stein* 1969a; *Poppele* and *Bowman* 1970; *Poppele* 1973; *Goodwin* et al. 1975; *Hunt* and *Ottoson* 1977; *Poppele* et al. 1979; *Hunt* and *Wilkinson* 1980; *Emonet-Dénand* et al. 1980c; *Baumann* et al. 1982; *Baumann* and *Hulliger*, to be published a). Such length dependence of small-movement sensitivity can be very pronounced, so that the sensitivity in a slack muscle may be less than a tenth of the value in a taut muscle. Thus, the size of the effect is of the same order of magnitude as the reduction of sensitivity provoked by powerful static fusimotor action (*Goodwin* et al. 1975; *Hulliger* et al. 1977a; *Cussons* et al. 1977; *Chen* and *Poppele* 1978) or by concomitant movements of large amplitude (Sect. 3.1.2). The length dependence of sensitivity can be largely accounted for by an increase, with length, of the polar muscle fibre stiffness (*Poppele* et al. 1979; cf. *Lännergren* 1971).

The extent of length dependence of small-movement sensitivity is reduced by both static and dynamic fusimotor action (*Goodwin* et al. 1975). Such linearization of incremental sensitivity was, in fact, already envisaged by *Kuffler* et al. (1951) as an important function of the fusimotor system. Thus, fusimotor neurones, if suitably activated, may help maintain small-movement sensitivity at an approximately constant level, in addition to maintaining spindle firing during muscle shortening (*Lennerstrand* and *Thoden* 1968c; cf. *Hunt* and *Kuffler* 1951a).

Harmonic Distortion. Responses to sinusoidal stretching of sufficiently large amplitude may strikingly deviate from the requisite sinusoidal waveform (*Poppele* and *Bowman* 1970; *Goodwin* et al. 1975; *Hulliger* et al. 1977a). This happens especially for passive afferents at frequencies of stretching above 5–10 Hz, when responses tend to consist of bursts of impulses occurring at fixed phases of the sinusoidal cycle (phase locking). In the extreme, responses to sinusoidal and rectangular stretches may look much alike, making decoding by any receiver virtually impossible.

Driving by Vibration. For frequencies of sinusoidal stretch above the afferent resting frequency, primaries – but not secondaries – can be driven to discharge a single impulse per cycle, provided that the amplitude of movement exceeds a critical but still small limit (*Bianconi* and *Van der Meulen* 1963; *Brown* et al. 1967a). This is a special case of phase locking (harmonic distortion), but it may also be looked at as the extreme manifestation of gain compression, since widely ranging amplitudes can then no longer be discriminated in the afferent response.

Velocity Response. In the small-movement linear range the velocity response of spindle afferents can be described by linear transfer functions (*Matthews* and *Stein* 1969a; *Poppele* and *Bowman* 1970; *Hasan* and *Houk* 1975a; *Hulliger* 1976). However, for larger movements linear description is no longer applicable (*Matthews* 1963; *Crowe* and *Matthews* 1964a; *Schäfer* 1973; *Cheney* and *Preston* 1976c; *Holm* et al. 1981; *Houk* et al. 1981). Instead, a power relationship between velocity (v) and response (r), $r = kv^n$, was more adequate. Yet the generality of such descriptions is limited, since k and n are not true constants. On the one hand, *Houk* et al. (1981) claimed that the exponent n, which is a measure of the degree of velocity dependence, indeed was invariant vis-a-vis variations of reflexly evoked fusimotor activity. On the other hand, *Schäfer* (1973) and *Holm* et al. (1981) found that it decreased during dynamic, and increased during static, γ-stimulation. Moreover, for passive primary afferents, n decreases with increasing muscle length (*Baumann* and *Hulliger,* to be published b).

Occlusion. During combined stimulation of a static and a dynamic fusimotor neurone, static action may completely occlude the effect provoked by dynamic action on its own (*Crowe* and *Matthews* 1964a,b; *Lennerstrand* 1968b; *Schäfer* 1974; *Hulliger* et al. 1977b; *Hulliger* and *Noth* 1979). Formally such occlusion is another example of violation of the superposition principle. As to the underlying receptor mechanisms, circumstantial evidence indicates that it may be a manifestation of competitive interaction of multiple pace-makers (*pace-maker switching*) (*Crowe* and *Matthews* 1964a; *Lennerstrand* 1968b; *Brokensha* and *Westbury* 1974; *Hulliger* et al. 1977c; *Hulliger* and *Noth* 1979; *Quick* et al. 1980).

After-effects of Fusimotor Activity. Stimulation of fusimotor neurones may leave the spindle in an altered state. This may last for several minutes unless it is abolished by large stretches (see Sect. 3.3). These after-effects may manifest themselves as an elevation of the resting discharge, as a marked enhancement of the initial burst at the beginning of a rapid stretch, or as an increase in small-movement sensitivity of Ia afferents during slow, concomitant movements of large amplitude (see Sect. 3.3).

In conclusion, the mammalian muscle spindle is a highly non-linear system. Some of the non-linearities, which are especially pronounced in passive afferents, are reduced by fusimotor action (gain compression, length dependence, harmonic distortion), yet others persist (velocity response), and some non-linearities are even introduced by fusimotor action (occlusion, after-effects).

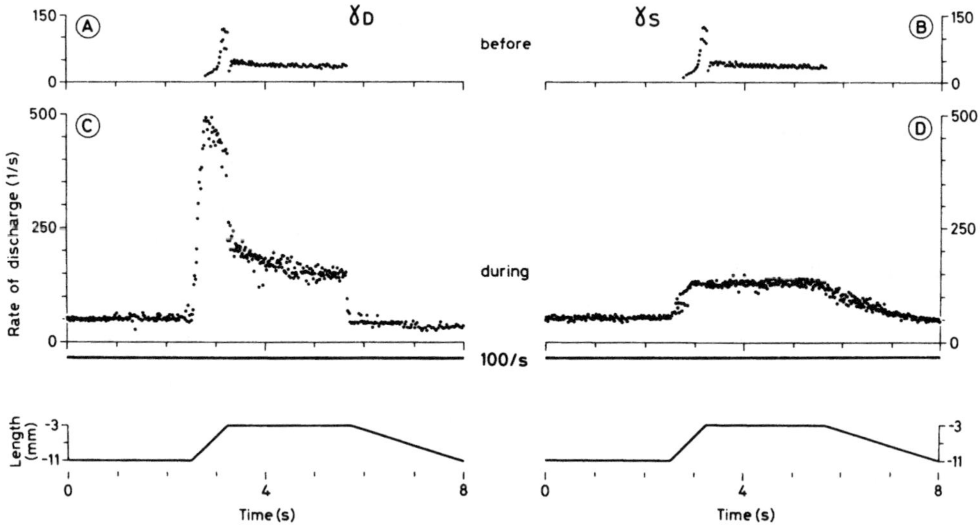

Fig. 8A–D. Action of dynamic and static fusimotor neurones on the response of a primary spindle afferent to large ramp-and-hold stretch. **A** and **B** Passive controls prior to fusimotor stimulation. **C** Response during tetanic stimulation of a dynamic γ-axon at 100/s. **D** Response during tetanic stimulation of a static γ-axon at 100/s. *Bottom traces,* trapezoidal time course of muscle stretch; amplitude 8 mm, starting from 11 mm below maximum physiological extension of soleus muscle. Same afferent unit in **A** to **D**. All responses displayed as single-sweep average frequency records. (Previously unpublished data of *Baumann* et al. 1982)

3.2 Static and Dynamic Fusimotor Neurones

3.2.1 Classical Properties

The original subdivision of fusimotor neurones into static and dynamic fibres (*Matthews* 1962) was based on functional rather than morphological criteria (see Sects. 2.1.3 and 2.2.1.4). This has continued to provide a valuable concept for the investigation of spindle function, and fusimotor neurones are still classified as static or dynamic mainly on the basis of their diminishing or enhancing action on the size of the dynamic response of Ia afferents to stretches applied to the parent muscle. In Fig. 8, the size of the dynamic index (i.e. the abrupt reduction in discharge rate at the end of the ramp stretch; *Crowe* and *Matthews* 1964a) was drastically increased during stimulation of a dynamic γ-fibre (cf. Fig. 8C with Fig. 8A). In contrast, static action reduced the size of the dynamic index to zero whilst exerting a powerful excitatory effect as seen in the increase in mean discharge rate (cf. Fig. 8D with Fig. 8B).

Ramp-and-hold, triangular and sinusoidal stretches have been widely used for the classification of fusimotor neurones (both γ- and β-fibres). Which particular function is chosen for the assessment of the dynamic response is irrelevant, provided that the amplitude of stretch is large enough (at least 2 mm peak-to-peak). For smaller stretches Ia dynamic sensitivity may – paradoxically – be reduced also by dynamic γ-action (*Goodwin* et al. 1975; *Hulliger* et al. 1977a; see also Fig. 10).

At first sight the *duality* of fusimotor action might seem trivial, since it is based on a binary assessment of a single parameter (increase or decrease of dynamic sensitivity). However, from the beginning the validity of the classification was supported by the observation that the action of a single γ-motoneurone on different spindles was either uniformly static or uniformly dynamic (*Crowe* and *Matthews* 1964b; *Brown* et al. 1965; *Bessou* et al. 1966; *Matthews* 1972, 1981b, reviews). On the whole this has since been confirmed whenever the action of single fusimotor fibres on several of its target spindles was examined (e.g. *Emonet-Dénand* et al. 1977b). This conforms with the *uniform specificity* seen in the innervation of intrafusal muscle fibres by functionally identified fusimotor neurones (see Sects. 2.1.4 and 2.2.1.4). Thus, the association between individual fusimotor neurones and one of the two types of fusimotor effects is unlikely to be due to mere chance. This has two important implications.

First, it suggests that during development an ordering principle must guide the diverging branches of fusimotor neurones to their specific intrafusal targets. Or else, fusimotor neurones, having established functional contacts with intrafusal fibres, might determine their differentiation into bag or chain fibres. The latter possibility seems unlikely in view of the relative unimportance of motor innervation for the normal development of intrafusal muscle fibres (see Sect. 2.2.1.4, motor terminals). Beyond this we have no certain insight into the mechanisms underlying these guiding principles, apart from detailed descriptive knowledge of the morphological events during ontogeny (*Zelená* and *Hnik* 1963; *Zelená* 1964; *Landon* 1972; *Milburn* 1973; *Butler* 1980; *Kozeka* and *Ontell* 1981; *Barker* and *Milburn* 1982; see also Table 1).

Second, individual fusimotor neurones, rather than their interaction with the target spindles, seem to possess a specific static or dynamic quality. This quality might therefore be accessible to selective *neural control* by peripheral reflexes or descending pathways from higher motor centres (see Sect. 4).

The responses shown in Fig. 8 are not quite classical, in that during the slow release of stretch tonic dynamic action was strong enough to prevent the afferent from falling silent (cf. Fig. 8C with Fig. 8A). However, from the outset it was a distinguishing feature of dynamic action that its powerful action during muscle *lengthening* contrasted with its unimpressive excitatory effects during *shortening*, permitting afferent silence during slow release. Static fusimotor fibres, on the other hand, were normally found to maintain afferent discharge during muscle shortening (see *Crowe* and

Matthews 1964b; *Lennerstrand* and *Thoden* 1968a,b,c; also *Emonet-Dénand* et al. 1977b). The slightly non-classical behaviour of the dynamic axon of Fig. 8 may be taken as a harbinger of the more recent recognition that the duality of the subdivision into two types of fusimotor neurone was not absolute (see Sect. 3.2.2).

Secondary afferents are normally operated only by static fusimotor neurones (*Appelberg* et al. 1966). Typically, static fibres exert pronounced excitatory effects, and they may increase static-position sensitivity (*Lennerstrand* and *Thoden* 1968b; *Brown* et al. 1969b; *Jami* and *Petit* 1978), but this is accompanied by smaller effects on dynamic sensitivity than with the primary afferents. Since the dynamic responsiveness of passive secondary afferents is low to start with, it is not surprising that in absolute terms any reduction in dynamic sensitivity tends to be small. Sometimes fusimotor action on secondary afferents may even slightly increase the size of their dynamic response (*Brown* et al. 1967b; *Cheney* and *Preston* 1976c; *Cussons* et al. 1977; *Jami* and *Petit* 1981).

3.2.2 Range of Fusimotor Effects

Among the effects exerted by different fusimotor neurones, there is considerable variability. Both the size of the excitatory effect (change in static firing or fusimotor bias) and the amount of change in dynamic sensitivity may span a wide range. This was manifest in the first scatter diagrams relating these two parameters (*Crowe* and *Matthews* 1964b; *Brown* et al. 1965). *Emonet-Dénand* et al. (1977b) reassessed the degree of purity of a large sample of fusimotor effects, relying on a number of criteria such as the effect on dynamic sensitivity and the occurrence of afferent silence during muscle shortening. In order to account for the range of fusimotor effects observed, these were allocated to one out of six categories, graded between pure dynamic (I) and pure static (VI) action. In retrospect the original binary classification still came close to the truth, since two-thirds of the effects were unambiguous (either static or dynamic). The remaining intermediate effects appeared as weighted mixtures of the two extremes, but predominantly static fibres showed a lesser degree of consistency in the type of effects on their target spindles than predominantly dynamic fibres.

At the time, these findings agreed with the apparently lesser degree of specificity of static innervation (involving also bag_1 fibres). However, given the recent consensus on static selectivity, this no longer seems likely (*Barker* and *Stacey* 1981; *Boyd* 1981a; see Sect. 2.1.4).

The greater variability of ***static effects*** might reflect inhomogeneities of mechanical and contractile properties (see Table 2) among the intrafusal fibres supplied, especially if the innervation profiles (for bag_2 and chain fibres) were not the same for all the spindles operated by a single static axon. For peroneus brevis (*Emonet-Dénand* et al. 1977b), data bearing on this particular possibility are not available. Yet for tenuissimus, observations on isolated spindles have indeed shown widely ranging innervation profiles

(*Boyd* and *Ward* 1975; *Boyd* et al. 1977). However, non-selective static innervation (bag_2 *and* chain) is rare (10%–15% of the γ_S-Ia interactions; deduced from *Boyd* et al. 1977). Thus, the majority of static axons are selective and consistent in innervating either bag_2 or chain fibres only (*Boyd* et al. 1977; *Boyd* and *Ward* 1982; *Boyd* et al. 1983b).

In view of such selectivity and given that bag_2 fibres might be selectively controlled from the motor cortex (*Gladden* and *McWilliam* 1977; *Gladden* 1981), *Boyd* and co-workers have suggested that there might be *two distinct classes of static γ-motoneurones* (see also *Boyd* and *Ward* 1982; *Arbuthnott* et al. 1982; *Boyd* et al. 1983b). This proposal agrees with the classification of *Emonet-Dénand* et al. (1977b) in that static fusimotor effects are not uniform, but it remains to be seen to what extent the two schemes agree in detail. This might not be the case if the selectivity of static innervation seen in tenuissimus were a feature of that muscle only.

3.2.3 Stength and Speed of Fusimotor Action

Both the strength of fusimotor action, as measured by the increase in afferent discharge rate (bias) at constant muscle length, and the speed of its manifestation at the onset of activation are important parameters for any consideration of the role of fusimotor neurones in motor control.

Strength. In the earliest investigations static and dynamic γ-motoneurones were equally effective in biasing primary afferents at short muscle length (*Matthews* 1962; *Crowe* and *Matthews* 1964b; *Brown* et al. 1965). However, in more recent measurements at longer muscle length static axons were consistently about twice as powerful in exciting Ia afferents as dynamic axons (*Lennerstand* and *Thoden* 1968a,b; *Andersson* et al. 1968b; *Lewis* and *Proske* 1972; *Cheney* and *Preston* 1976c; *Emonet-Dénand* et al. 1977b, Hulliger 1979). The difference between these two sets of data may simply reflect the different actions on position sensitivity, which tends to be increased by static and decreased by dynamic fibres (*Brown* et al. 1969b; *Cheney* and *Preston* 1976c). Thus, all in all static γ-motoneurones are clearly more powerful in biasing Ia afferents (cf. also Table 2, extension of sensory spirals).

This difference in excitatory strength is particularly pronounced during rate-modulated stimulation (Fig. 9) when the discharge at two different levels of activation is compared (instead of assessing the effects of tonic stimulation against the afferent's resting discharge, as above). On average the static/dynamic ratio in excitatory strength was in excess of a factor of four (*Hulliger* 1979).

Functionally the difference in excitatory strength seems all the more significant, as individual spindles are normally supplied with two to three times more static than dynamic fusimotor fibres (see Sect. 2.2.1.4). Thus, the static fusimotor system appears to be a particularly powerful tool for the control of the overall discharge level of spindle afferents. In contrast, dynamic fusimotor neurones seem to be well suited for the control of the dynamic sensitivity by virtue of their powerful sensitivity-enhancing action.

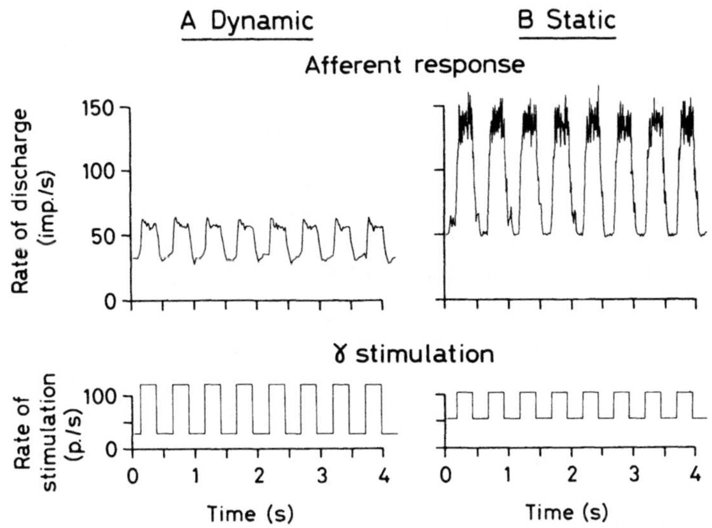

Fig. 9A,B. Responses of a primary spindle afferent unit to rectangular stimulation of a single dynamic (**A**) and a single static (**B**) fusimotor fibre. *Upper graphs,* time course of instantaneous rate of discharge; *lower graphs,* schematic illustration of the time course of stimulation rate. Note that the responses to static stimulation are about three times as large as with dynamic stimulation in spite of the smaller modulation of stimulation rate (static, 50/s; dynamic, 90/s). (From *Hulliger* 1981, with permission)

The weaker excitatory strength of dynamic γ-axons is a feature of measurements taken at fixed muscle length. In contrast, a few or even a single impulse in a dynamic γ- or β-fibre – if applied during dynamic stretch – may powerfully excite Ia afferents (*Emonet-Dénand* and *Laporte* 1981), probably by increasing polar stiffness in bag_1 fibres (*Matthews* 1981a) and perhaps by stretch activation (cf. Sect. 2.2.1.2). Whatever the mechanism, the observation is in line with the powerful sensitivity-enhancing action of dynamic fibres, which often is already pronounced during low-level tetanic activation (e.g. Fig. 12C).

Speed. Dynamic fusimotor action on primary afferents has long been held to be considerably slower than static action (*Westbury* 1964; *Brown* and *Matthews* 1966; *Bessou* et al. 1968; *Boyd* 1976a,b; *Boyd* et al. 1977). This seemed a useful concept to account for the striking difference in their effects on primary afferents' dynamic sensitivity. By virtue of their high cross-bridge turnover rate, *fast-contracting* fibres should be capable of absorbing a major fraction of imposed dynamic length changes without generating viscous (i.e. dynamic) tension transients, since the bridges would be attached for too short a time to be sufficiently strained. In contrast, *slowly contracting* fibres with a much lower turnover rate would generate viscous tension, since the cross-bridges would be attached long enough to be appreciably strained during imposed length changes. Thus the differences in dynamic response as seen, for example, during ramp stretches could be explained by differences in contraction speed, provided that these were sufficiently large.

This requirement seems to be met in amphibia, where the contraction speeds of slow and fast skeletal muscle fibres may differ by more than a factor of ten (*Kuffler* and *Vaughan-Williams* 1953), and where pronounced visco-elastic tension transients during stretching are encountered on activation of slow but not of fast fibres (*Brown* 1971b). Amphibian muscle is in part innervated by either fast or slow β-motoneurones, which exert static and dynamic fusimotor effects (*Matthews* and *Westbury* 1965; *Brown* 1971a). If the mechanical properties of intra- and extrafusal fibres of a given β-motoneurone were similar, then dynamic action would indeed by mediated by slowly contracting fibres and static action by much faster contracting fibres.

For the mammalian spindle the difference in contraction speed between bag_1 fibres and bag_2 or chain fibres seems too small to account for the large-scale difference between dynamic and static fusimotor action on Ia sensitivity. At the intrafusal level, relatively small differences in contraction time were found, with static action being about twice as fast as dynamic action (*Bessou* and *Pagès* 1975; *Boyd* 1976a,b; *Boyd* et al. 1977). However, quantitative measurements of the speed of fusimotor action on primary afferents have failed to reveal a corresponding difference (*Andersson* et al. 1968b; *Chen* and *Poppele* 1978; *Hulliger* 1979). This may be seen in the example of Fig. 9, showing that the speed of a Ia response to step changes in fusimotor drive was much the same with dynamic (Fig. 9A) and static (Fig. 9B) γ-action, whilst the amplitude of response was strikingly different (see above). The reason for this difference in speed between intrafusal contraction and afferent response is not known. Yet it seems significant functionally that both types of fusimotor neurone are capable of exerting their specific action on primary afferents equally quickly.

Internal Spindle Mechanisms. Static fusimotor effects are non-controversial and widely attributed to the activation of fast-contracting muscle fibres. The reduction of dynamic sensitivity of spindle afferents is then explicable in terms of fast cross-bridge kinetics (see above). This view is based on the observation of apparently high tetanic fusion rates of those intrafusal muscle fibres which are operated by static γ-axons. During tetanic stimulation twitchlike events, causing modulation in spindle afferent discharge, seem to occur and are revealed in frequencygrams (*Bessou* et al. 1968; *Emonet-Dénand* and *Laporte* 1969) or in post-stimulus histograms (*Goodwin* et al. 1975; *Cussons* et al. 1977; *Emonet-Dénand* et al. 1977a; *Hulliger* 1979). These stimulus-locked events may even be so strong as to elicit 1:1 driving of Ia discharge at the rate of static fusimotor stimulation (*Crowe* and *Matthews* 1964b; *Brown* et al. 1965; *Boyd* and *Ward* 1982; *Boyd* et al. 1983a). However, whilst such driving on its own may powerfully reduce sensitivity to stretching (by curtailing discharge modulation; see *Chen* and *Poppele* 1973; *Boyd* et al. 1983a) it is not a necessary feature of static action, since non-driving static fibres may induce equally pronounced reduction in sensitivity (*Goodwin* et al. 1975, 1976; *Hulliger* et al. 1977a; *Hulliger* 1979). This ties in with the observation of *Boyd* (1981b) that among the intrafusal fibres operated by static axons only the chain, but not the bag_2 fibres show fast twitch contractions.

The mechanisms of dynamic fusimotor action are more enigmatic. Uniformly slow cross-bridge kinetics can no longer be invoked (see *Hulliger* 1979). The recent proposal of *Matthews* (1981a) that non-uniformities in contraction speed along the bag_1 fibre might be responsible for the 'viscous' properties of dynamic action would have to be verified by local measurements of – preferably neutral (see Sect. 2.2.1.2) – ATPase activity or of contraction speed. Another view is that stretch activation of the bag_1

fibre, facilitated by dynamic fusimotor action, is at the heart of the matter (see *Poppele* and *Quick* 1981; also *Proske* 1975, and Sect. 2.2.1.2). Such stretch activation would manifest itself in the develeopment of viscous force during stretching. Moreover – in line with *Matthews'* hypothesis (1981a) – its effect on dynamic sensitivity of primary afferents might be potentiated if it occurred non-uniformly along the bag_1 fibre. However, this hypothesis too requires verification in recordings of local contractile events as obtained, for instance, from measurements of time course of sarcomere spacings.

3.2.4 Action on Sensitivity to Sinusoidal Movements of Widely Ranging Amplitude and Frequency

If the frequency response of spindle afferents could be described in terms of higher-order linear high-pass filters, then fusimotor action might affect either the gain factor or the various time constants. Only in the latter case would changes in overall gain (sensitivity) be accompanied by changes in phase for a given frequency. However, the non-linearities of the spindle have defied such simple modelling (see *Hasan* 1983). Notwithstanding such theoretical expectations, frequency-response analysis has clearly shown that in the physiological range of movements, corresponding to sinusoidal frequencies up to about 20 Hz, fusimotor action controls the gain of response rather than its phase, relative to the driving function. The relevant findings were drawn from a limited number of studies, which for convenience are cited only once: *Matthews* and *Stein* (1969a), *Rosenthal* et al.(1970), *Goodwin* et al. (1975, 1976), *Hulliger* et al. (1977a,b), *Cussons* et al. (1977), *Chen* and *Poppele* (1973, 1978).

The *gain control* provided by fusimotor neurones is straightforward for static efferents, as long as they are activated at high rates (but cf. Sect. 3.4). Compared with passive control responses, static fibres then reduce sensitivity over a wide range of movement amplitudes. This effect is large for primaries but smaller for secondaries. Dynamic action, when studied at constant mean muscle length, is somewhat unorthodox. For very small movements it consists of a reduction of Ia sensitivity (see Fig. 10), and only for movements of intermediate or large amplitude does it lead to the familiar and pronounced enhancement of Ia sensitivity (at 1 Hz above 200 μm). However, the manifestation of this paradoxical effect is restricted to a narrow range of motor paradigms (such as postural tasks where mean muscle length may be approximately constant). Preliminary evidence indicates that in the more general case when small stretches are superimposed on large movements and when passive Ia sensitivity tends to be rather low (see Sect. 3.1.2), dynamic action also leads to an augmentation of primary afferent responses to even the smallest movements (cf. Fig. 12C with Fig. 12A; also *Baumann* et al., unpublished).

During natural movements purely dynamic or purely static fusimotor outflow to the muscles concerned may occur only rarely. Various com-

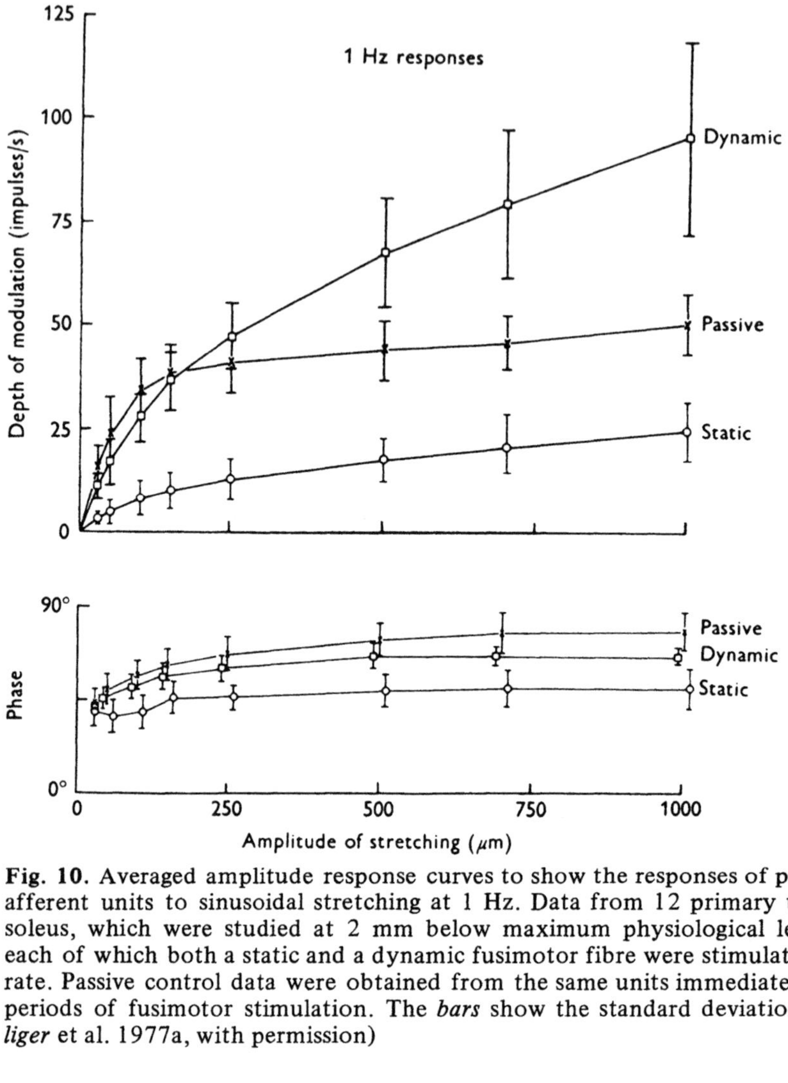

Fig. 10. Averaged amplitude response curves to show the responses of primary spindle afferent units to sinusoidal stretching at 1 Hz. Data from 12 primary units from cat soleus, which were studied at 2 mm below maximum physiological length, and for each of which both a static and a dynamic fusimotor fibre were stimulated at constant rate. Passive control data were obtained from the same units immediately prior to the periods of fusimotor stimulation. The *bars* show the standard deviation. (From *Hulliger* et al. 1977a, with permission)

binations of the two, differing in the relative weight given to each one, probably are the rule. This seems all the more likely in view of the fixed amounts of fusimotor drive provided by dynamic and static β-fibres once they are recruited to fulfill their extrafusal role. Thus, the perhaps most functionally relevant insights from frequency response studies came from the investigation of *combined static and dynamic fusimotor action on the primary afferent.*

Generally, sensitivity was graded between a maximum set by dynamic action on its own and a minimum set by static action on its own, the precise level of response being determined by the relative strength of the two types of action. However, for small movements dynamic action was completely occluded by even small amounts of static drive. Therefore, if sizable values of sensitivity are required to monitor small disturbances, dynamic fusimotor neurones would have to be activated on their own, to the exclusion even of static β-fibres (see below). Thus for a fixed mechanical stimulus, spindle afferent feedback need not be invariant. Instead, its parameters may be flexibly adjusted, provided that static and dynamic fusimotor neurones can independently be controlled by descending pathways. The fusimotor system as a whole may therefore be viewed as providing *parameter control* (see *Matthews* 1981b,c).

Fusimotor effects on *phase* are small and probably functionally insignificant when compared with the often large effects on sensitivity. Both static and dynamic action cause a slight phase lag in Ia responses to movements of widely ranging amplitude (Fig. 10) and frequency. Static action on secondaries does not alter phase at all.

Such absence of phase control is of interest in relation to technical feedback-control systems. In these, feedback gain is often controlled by varying the amount of velocity feedback for invariant position feedback, in order to stabilize an otherwise oscillatory system. Similarly, biological motor-control systems which include sensory feedback loops may show an oscillatory tendency because of inertial loads and conduction and activation delays. Looked at from this angle, it is remarkable that the fusimotor system is apparently not designed to provide velocity or phase control of the feedback signals from the muscle spindle in order to damp any inherent oscillatory properties, for instance, of the stretch reflex.

3.3 After-effects of Fusimotor Activity

For high stimulation rates, onset and termination of fusimotor excitation of spindle afferents are nearly instantaneous (see, e.g. *Hulliger* 1979; cf. Sect. 3.2.3). Yet after stimulation the spindle may be left in an altered state of responsiveness (fusimotor after-effects). It is common experience that long-lasting stimulation, especially at high rates, is often followed by an early depression of resting discharge or of responses to stretch (see *Kuffler* et al. 1951; *Hutton* et al. 1973; *Durkovic* 1976). On the other hand, post-excitatory facilitation has also been observed, either immediately after brief stimulation (*Kuffler* et al. 1951; *Hunt* and *Kuffler* 1951a; *Kidd* 1964; *Hutton* et al. 1973, 1975; *Smith* et al. 1974), or delayed beyond the period of depression (*Kuffler* et al. 1951; *Hutton* et al. 1973; *Durkovic* 1976). Common to most of these observations was that they were long lasting (up to 1 h, *Hutton* et al. 1975) and that they were abolished by stretches of sufficiently large amplitude.

Facilitatory after-effects have also been demonstrated in the augmentation of the initial Ia response to fast and large stretch ('initial burst') upon stimulation of both static and dynamic γ-fibres (*Brown* et al. 1969a; *Durkovic* 1976; *Gregory* et al. 1977). These effects were of interest for con-

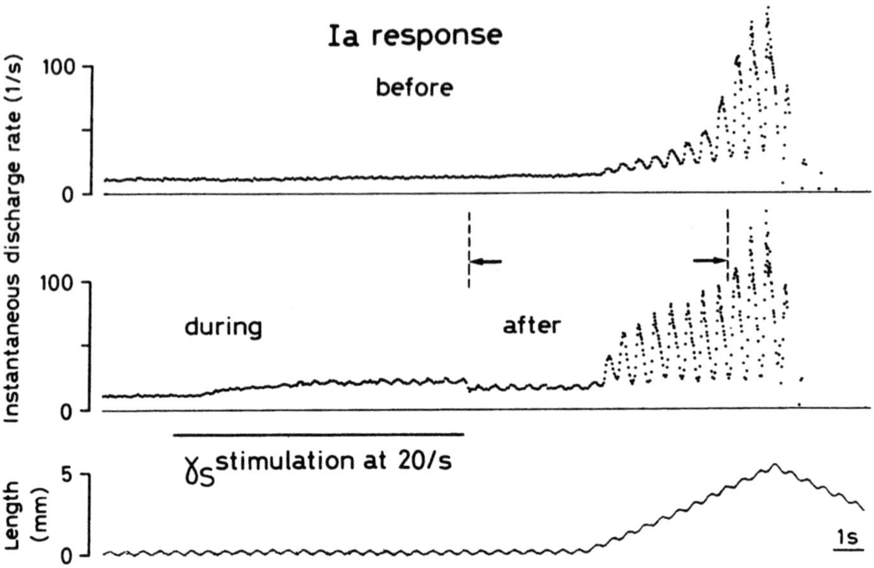

Fig. 11. Sensitivity-enhancing after-effects following low rate stimulation (20/s) of a static γ-fibre at resting length (0 mm, corresponding to ankle joint angle of 90°). Traces of instantaneous frequency of Ia discharge. *Top,* Passive control response of Ia afferent. *Middle,* Stimulation and after-effects in Ia response to sinusoidal stretches at short resting length and during large triangle stretch. *Bottom,* Time course of muscle stretch. (From *Baumann* et al. 1983b, with permission)

siderations of receptor mechanisms and mostly attributed to the persistence of stuck cross-bridges between intrafusal actin and myosin filaments (cf. 'small-movement sensitivity', Sect. 3.1.2). This was in keeping with the observation that, again, such facilitation lasted for minutes unless it was abolished by a subsequent large stretch (*Brown* et al. 1969a).

γ-Mediated post-excitatory facilitation of Ia afferents probably also played an important role in the enhanced sensory discharge from muscles which was observed after stimulation of entire ventral roots or after reflex-induced muscle contraction ('post-contraction sensory discharge', *Hnik* et al. 1970; *Hutton* et al. 1973, 1975; *Smith* et al. 1974). However, contributions from non-proprioceptive afferents or changes in the extrafusal mechanical or chemical environment could not be ruled out entirely nor could they be assessed quantitatively (see *Hutton* et al. 1973). Nevertheless it merits attention that such post-contraction facilitation not only increased muscle afferent (and Ia) resting discharge, especially at short length, but that it also enhanced dynamic response during slow stretch (*Smith* et al. 1974). However, this could not simply be ascribed to persisting effects of dynamic fusimotor activity, since *Brown* et al. (1969a) had described a similar after-effect of static γ-stimulation.

These observations have recently been confirmed and further extended. Sizable after-effects on *dynamic sensitivity,* occurring over a wide range of muscle lengths, were found following stimulation of both dynamic or static γ-fibres at low rates (mostly between 10 and 50/s; *Baumann* et al. 1982, 1983a,b and unpublished). These effects manifested themselves as

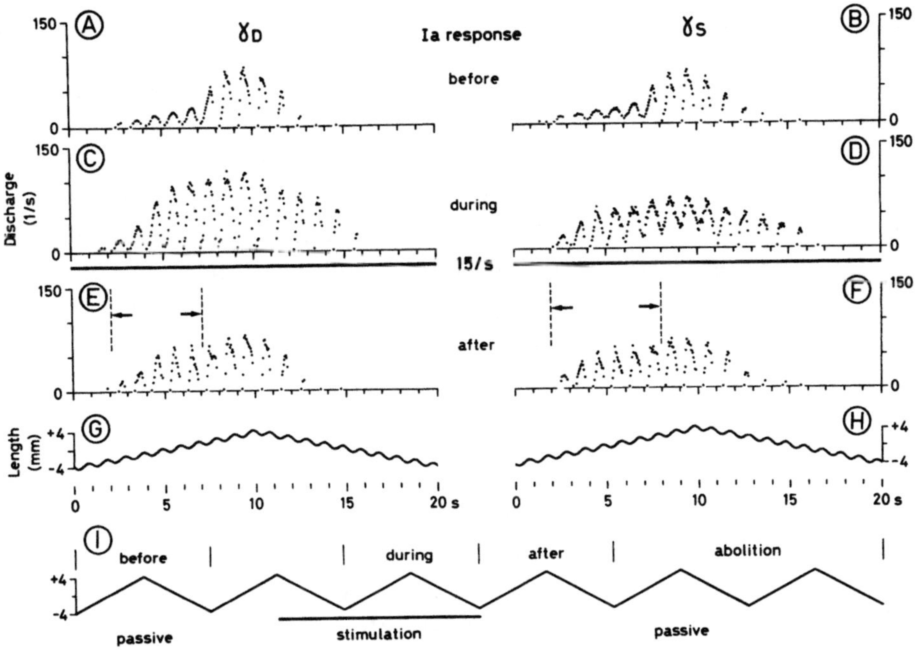

Fig. 12A–I. Dynamic and static fusimotor action on response to small sinusoidal stretches, during and after low rate stimulation of single γ-fibres and during ongoing triangular movements of large amplitude. **A, B** Passive control responses to combined sinusoidal and triangular movement prior to stimulation (see **I**). **C** Response during tonic γ_D-stimulation at 15/s. **D** Tonic γ_S-stimulation at 15/s. **E, F** After-effects marked by *arrows* and *vertical dashed lines* (compare with **A** and **B**). **G, H** Schematic display of muscle stretch (triangle 4 mm, centred around intermediate muscle length, 0 mm, corresponding to 90° ankle joint angle, sinusoids 0.5 mm; half peak-to-peak amplitudes). **I** Schematic illustration of test sequence. Single-sweep average frequency records; same afferent and efferent fibres as in Fig. 8. (Modified from *Baumann* et al. 1982, with permission)

an often pronounced increase in the responses to small sinusoidal stretches superimposed on large triangular movements (see Figs. 11; 12E, F). Such enhancement of sensitivity was especially prominent at short and intermediate muscle length. These effects were most effectively provoked by γ-stimulation either at short and fixed muscle length or during ongoing movement, the latter – interestingly – only during shortening but not during lengthening movements (*Baumann* et al. 1982). Finally, in agreement with the earlier studies (above), such after-effects could be provoked by short trains of γ-impulses; they lasted for several minutes, and they were abolished by a single stretch of large amplitude.

The mechanisms responsible for such widely ranging after-effects remain to be elucidated. For those effects provoked by only a few impulses in functionally single γ-fibres, changes in the extrafusal chemical or mechanical environment seem unlikely (cf. also *Hutton* et al. 1973). If stuck intrafusal cross-bridges were to be responsible, these would have to sustain much larger stretches than was hitherto thought possible (see *Hill* 1968; *Goodwin* et al. 1975; *Hasan* and *Houk* 1975b; *Hunt* and *Wilkinson* 1980).

By extending fusimotor action beyond the period of fusimotor activity, such after-effects might be functionally meaningful. They may improve spindle performance at short muscle length by increasing resting discharge or by enhancing otherwise poor small-movement sensitivity (see Sect. 3.1.3). However, if the latter mattered, this would be at the expense of the distinction between static and dynamic neurones, since persisting enhancement of sensitivity may be provoked by both of them.

3.4 Static Fusimotor Action at Low Stimulation Rates

At present the rates of discharge of γ-motoneurones during natural movements are not known, since direct recordings from these fibres have not so far been achieved. If the discharge rates of α-motoneurones, which rarely exceed 40/s, may serve as a standard of reference, then most experimental work in anaesthetized preparations has until now been performed with unnaturally high stimulation rates, ranging from 70 to about 150/s. At these rates the fusimotor effects of single axons are pronounced and, mostly, clear-cut. Their use may eventually be warranted if they turn out to be adequate models of simultaneous low-rate activation of the entire fusimotor supply of a single spindle.

The issue of low-rate fusimotor action is by no means trivial, since activation of fusimotor neurones at low stimulation rates may give paradoxical effects. Thus, static γ-fibres (unambiguously classified using high stimulation rates) may, during low-rate activation, increase rather than decrease the dynamic sensitivity of primary spindle afferents. The first report on this phenomenon (*Emonet-Dénand* et al. 1972) has not gained general acceptance, owing to limitations of analysis. However, a reinvestigation of this issue has confirmed these findings (*Hulliger* et al. 1983). A closely related phenomenon is illustrated in Fig. 12, where the responses to sinusoids during slow stretch were clearly augmented at short length when a static axon was stimulated at 15/s (cf. Fig. 12D with Fig. 12B). This *'static paradoxon'* is only seen at short or intermediate muscle length, but there in about two thirds of the γ_S-Ia interactions (*Hulliger* et al. 1983). However, the most relevant and still unanswered question is whether this phenomenon persists on simultaneous low-rate activation of several 'paradoxical' static axons.

3.5 Role of β-Fusimotor Neurones

β-Innervation is widespread (see Sect. 2.1.5) and β-action can be powerful (see, e.g. *Bessou* et al. 1965; *Emonet-Dénand* et al. 1975, 1980a; *Post* et al. 1980; *Emonet-Dénand* and *Laporte* 1983). During voluntary movement β-action is therefore likely to contribute a substantial component to overall fusimotor outflow which, moreover, is strictly linked to skeletomotor outflow.

Slow β-fibres exert dynamic fusimotor effects and fast β-fibres exert static fusimotor effects (see Sect. 2.1.5). Conceivably, the descending and reflex control of β-fibres is substantially different from that of ordinary, i.e. purely skeletomotor, α-fibres. However, there is as yet no evidence to support this notion (see *Laporte* et al. 1981). Therefore the recruitment of β-fibres during a wide range of movements ought, at least partly, to be governed by the 'size principle' (*Henneman* et al. 1965a,b; see also *Henneman* and *Mendell* 1981). According to this rule, small α-motoneurones are normally activated before large α-motoneurones. This holds true especially for isometric contractions and slow isotonic or auxotonic movements (see *Burke* and *Edgerton* 1975; *Buchthal* and *Schmalbruch* 1980). Therefore, in finely graded movements involving small contraction forces, dynamic β-action should come into play before static β-action (see also *Matthews* 1981a). Conceivably, such automatic activation of dynamic fusimotor neurones may be advantageous, particularly for skilled movements when the generation of small forces has to be combined with that of finely graded movements.

However, the β-fusimotor automatism need not always be implemented in isolation, since at the same time as dynamic β-fibres are recruited independent activation of γ-fusimotor neurones may also take place. If this involved a substantial static contribution, the dynamic β-action on primary afferents might largely be occluded, especially in responses to small movements such as unwanted disturbances (see Sect. 3.2.4, combined stimulation). On the other hand, in combination with dynamic γ-action, the β-effect on Ia dynamic sensitivity might play an important facilitatory role.

3.6 Future Directions

The progress during the past two decades in the understanding of the functional properties of the mammalian muscle spindle may largely be attributed to the use of increasingly refined techniques and to the pursuit of carefully controlled experiments. Yet what is easy to control and convenient to analyse is not necessarily what prevails under natural conditions.

For instance, studies on small-movement sensitivity have mostly been performed on maximally stretched muscles and γ-activation rates have often been from 100/s upwards. Yet during natural movement such extremes are probably only rarely reached. The following aspects should therefore merit attention.

Length and Movement. To investigate spindle function within the range of muscle length where natural movements also occur is a mere calibration task, since the requisite information is available in detail (*Goslow* et al. 1973). Imitation of natural stepping, walking or swaying movements is also within reach of present techniques (see Sect. 5.2.4).

Rates and Patterns of γ-Stimulation. The use of relatively low (10–50/s) and probably physiological stimulation rates needs hardly be further justified. Since the size of the fusimotor effects elicited by such stimulations regimes is bound to be moderate, this should almost automatically lead to the important investigation of multiple fusimotor action.

Concomitant Variation of Muscle Length and Fusimotor Drive. Voluntary movements are brought about by variations in skeletomotor activity. One need not be an advocate of tight α-γ linkage to envisage the possibility that movements might be accompanied by fluctuations in fusimotor outflow. The dearth of experimental data on the effects on concomitant variation of muscle length and fusimotor drive (*Lennerstrand* and *Thoden* 1968c; *Appenteng* et al. 1982b) can only be a challenge for future research.

4 Central and Reflex Actions on γ-Motoneurones

The aim of this section is to give an overview emphasizing both similarities and differences in the control of α- and γ-motoneurones on the one hand, and of static and dynamic γ-motoneurones on the other hand. This bears on the general questions whether, and to what extent, these three types of motoneurone are controlled in parallel or independently of each other. For more detailed accounts of the bewildering variety of observations on this subject, recent reviews and monographs may be consulted (*Baldissera* et al. 1981; *Grillner* 1969b, 1976; *Matthews* 1972; *Murthy* 1978).

The activity of γ-motoneurones is determined by three types of neuronal imput: first, by pathways from *supraspinal* motor centres; second, by *propriospinal* circuitry; and third, by exteroceptive or proprioceptive *reflexes.* This classification may well be artificial, yet it reflects quite adequately the current state of experimental techniques. Consideration of proprio-spinal mechanisms is restricted to recurrent inhibition (Sect. 4.3), since for γ-motoneurones little else is known about such circuitry. This is in marked contrast with the detailed understanding of the role of proprio-spinal mechanisms for the control of α-motoneurones (see *Lundberg* 1979a; *Baldissera* et al. 1981).

4.1 Methods of Investigation

At present, no single method is available that could provide comprehensive information on synaptology and selectivity of a particular pathway, and on strength and type of fusimotor effects elicited. At best, combinations of complementary techniques can be used; yet, even so, the interpretation of the data obtained is often ambiguous. In the following, the most commonly used techniques are briefly but critically reviewed.

4.1.1 Intracellular Recordings

The strength of intracellular recording lies in the precision of information that may be obtained on the synaptic organization of input pathways. Its main limitation for γ-motoneurones is the mostly limited size of the sample, which is attributable to the technical difficulties in recording from small cells. Intracellular recordings are further limited, as they do not permit a direct classification of γ-cells as either static or dynamic, and as they tend to be biased towards analysis of short-latency effects.

4.1.2 Extracellular Recordings

Extracellular recordings from single γ-motoneurones may be obtained from the vicinity of the cell soma or from isolated filaments (in ventral roots or muscle nerves). The main advantage of these techniques is that they are less demanding technically and therefore they tend to yield larger samples of γ-cells.

The limitations are that they rarely permit unambiguous deductions about the synaptic connectivities of the inputs under investigation and that, as for intracellular recordings, it is not normally possible to directly classify the γ-cells recorded from. Perhaps this difficulty may be overcome by studying γ-axons which are initially preserved in continuity in fine ventral root filaments (*Noth* and *Thilmann* 1980; *Noth* 1981). Central and reflex actions can then be investigated after cutting the filament by recording from its central end. Theoretically, stimulation of the peripheral stump of the same axon can then be used for its classification. However, the latter task has yet to be achieved (*J. Noth*, personal communication).

Common to all direct recordings from γ-motoneurones are the difficulties encountered in any attempt to classify them according to the original criteria (Sect. 3.2.1). Moreover, whilst the size of the responses of individual cells can be measured, the size of the fusimotor effects provoked in spindle afferents can only be estimated.

4.1.3 Classification of γ-Motoneurones as Static or Dynamic

Static and dynamic γ-motoneurones can only be differentiated by their fusimotor effects, and not by their conduction velocity (Sect. 2.2.1.4). During recordings from γ-cells this would only be possible if the cell under investigation could be selectively stimulated and if, at the same time, Ia responses to muscle stretching could be recorded from one of the target spindles of the γ-cell. This is technically difficult and has not so far been achieved. Partly these difficulties arise from the sizable matching problem in muscles which are richly supplied with spindles and fusimotor neurones. Perhaps the task could be cut down to practicable size by studying partially denervated muscles, much in line with the technique which has been so successfully used for the investigation of β-innervation (Sect. 2.1.5).

For want of something better, two approximative and indirect methods of γ-cell classification have been used with some success.

4.1.3.1 Stimulation in the Mesencephalon. *Appelberg* and his co-workers developed a mesencephalic stimulation technique to differentiate between static and dynamic γ-cells (see *Appelberg* 1981, review). By stimulation in a circumscribed area dorsal and caudal to the red nucleus, dynamic γ-motoneurones could apparently be selectively influenced (either excited or inhibited). This was deduced from observations of the appropriate effects on the dynamic sensitivity of primary spindle afferents (*Appelberg* and *Emonet-Dénand* 1965; *Appelberg* and *Molander* 1967). Static fusimotor effects were not noticeable with such juxtarubral stimulation. The region concerned has since been referred to as the mesencephalic area for dynamic control (MesADC).

The principal advantages of this method are (apart from its obvious merit as an approximative tool for classification) its immediate reversibility and its applicability to both flexor and extensor motoneurone pools (*Appelberg* and *Emonet-Dénand* 1965; cf. DOPA, Sect. 4.1.3.2). Its principal limitation lies in the uncertainty concerning the degree of selectivity, since the original demonstration of the dynamic fusimotor effect was based on a small sample of primary (but not of secondary) spindle afferents (*Appelberg* and *Emonet-Dénand* 1965; *Appelberg* and *Molander* 1967). Clearly, this uncertainty could be removed by a systematic assessment of the effects elicited from MesADC (see *Appelberg* et al. 1981).

4.1.3.2 Use of DOPA. In spinal preparations, dynamic but not static γ-motoneurones are spontaneously active (*Alnaes* et al. 1964). However, for flexor γ-motoneurones this disposition can be reversed by the application of DOPA (*Bergmans* and *Grillner* 1968a, 1969; *Grillner* et al. 1967). This was deduced from the changes seen in spindle afferent responses to ramp stretches when after DOPA the dynamic sensitivity of primary afferents was decreased; further, primary and secondary afferents showed signs of tonic static drive (*Grillner* et al. 1967; *Bergmans* and *Grillner* 1968a). From the evidence above, it was concluded that DOPA, via a monoaminergic reticulospinal pathway (*Andén* et al. 1966a,b), acted reciprocally on static and dynamic γ-motoneurones by exciting the former and inhibiting the latter. Therefore, the administration of DOPA has been used to differentiate between static and dynamic flexor γ-motoneurones.

The limitations of this technique are appreciable. As with the stimulation of MesADC, the characterization of fusimotor action in spindle recordings was based on a small sample of afferents (*Bergmans* and *Grillner* 1968a), so that the degree of selectivity of this method can not be fully assessed either. Further, the reciprocity of effects holds up for flexor γ-motoneurones only. In extensor muscles both types of γ-cells were found to be excited by DOPA (*Grillner* 1969a). Moreover, the method cannot be used in other than spinal preparations because of large pharmacological side-effects (*Andén* et al. 1966a). Finally, the method is not quickly reversible, since the DOPA effects wear off only gradually.

4.1.4 Recording of Afferent Discharge

Central and reflex actions on γ-motoneurones can also be recorded indirectly by analysing their effects on spindle afferent responses to standard stretches. The main advantage of this approach, especially when it is quantitative, is that it permits classification of the fusimotor effects as static or dynamic, relying on the original criteria.

Moreover, the degree of contamination of one type of action by the other can also be estimated (see *Appelberg* et al. 1981). Further, the measures taken estimate total fusimotor outflow to the spindle at issue. For functional considerations this may be more relevant than quantitative data from single γ-motoneurones. And finally, the method is not biased towards an analysis of short-latency effects (as intracellular recordings are).

On the negative side, such recordings do not provide information on the synaptic organization of the pathways concerned, since reliable latency estimates cannot be obtained. Also, unless EMG is recorded with highest resolution, γ-action cannot be separated from β-action. Moreover, inhibitory effects can only be detected when fusimotor neurones discharge spontaneously. Finally, weak effects on one type of fusimotor neurone may go unnoticed when activity of the other type is dominant. This holds true especially for weak dynamic fusimotor effects when static action prevails (see Sect. 3.2.4).

4.2 Descending Control of γ-Motoneurones

4.2.1 Similarities and Differences Between α- and γ-Motoneurones

Soon after the exclusive fusimotor role of γ-motoneurones had been recognized it became clear that they could be powerfully excited or inhibited from various parts of the brain (*Granit* and *Kaada* 1952; *Granit* et al. 1952; *Eldred* et al. 1953). These observations have since been further extended. Among the structures which may contribute to descending control of γ- or at least fusimotor neurones (cf. Sect. 4.1.4), the following may be named (see also references above and below): *motor cortex* and pyramidal tract (*Mortimer* and *Akert* 1961; *Kato* et al. 1964; *Gilman* et al. 1971b), *cerebellum* (*Granit* et al. 1955; *Gilman* and *McDonald* 1967; *Gilman* 1968), ventrolateral nucleus of the *thalamus* (*Yanagisawa* et al. 1963; *Gilman* et al. 1971a), *reticular formation* (*Shimazu* et al. 1962a,b).

Initially the interest was centred around the question of whether α- and γ-motoneurones were regularly co-excited or co-inhibited from central structures as at the time seemed to be the case for peripheral reflexes (*Hunt* 1951; *Kobayashi* et al. 1952; *Eldred* and *Hagbarth* 1954; *Hunt* and *Paintal* 1958; see *Granit* 1955, p. 254, also *Hunt* and *Perl* 1960). It emerged that a tight parallelism, both in time and intensity, of α- and γ-activity was not the rule. As was the case with reflex actions, γ-responses to central stimulation were often manifest before α-responses (*Granit* and *Kaada* 1952; *Eldred* et al. 1953; *Granit* and *Holmgren* 1955; *Granit* 1955, review). This finding of a *lower γ-threshold* for central activation has since repeatedly been confirmed (see, e.g. *Laursen* and *Wiesendanger* 1966; *Fidone* and *Preston* 1969; *Grigg* and *Preston* 1971; *Vedel* 1975).

Irrespective of the underlying circuitry (separate pathways vs size-governed recruitment by a single pathway), large differences in threshold may from a functional point of view be looked at as a tool for graded differential control of α- and γ-motoneurones. If these differences were also present during normal movement, fusimotor neurones would be recruited before skeletomotor neurones. Particularly for slow and finely graded movements, this could be of considerable importance for parameter control of spindle afferent feedback (see Sect. 3.2.4, and *Matthews* 1981c).

The patterns of central activation of α- and γ-motoneurones range between closely knit coactivation and mutual independence. *Coactivation* has been described for the following descending systems.

First, monosynaptic excitation of both α- and γ-motoneurones to extensor muscles has been found on stimulation of the lateral vestibular nucleus, from which axons descend to the cord via the *vestibulospinal tract* (*Grillner* et al. 1969; *Grillner* 1969a). This pathway seems to act on static γ-motoneurones relatively selectively (*Bergmans* and *Grillner* 1968b; *Grillner* et al. 1969; see below).

Second, monosynaptic coactivation of flexor α- and γ-motoneurones was encountered when fibres in the *medial longitudinal fascicle,* probably originating from the pontine *reticular formation,* were activated (*Grillner* et al. 1969). Again, static γ-motoneurones seemed predominantly involved (*Bergmans* and *Grillner* 1968b).

Third, the *rubrospinal tract* has been found to act reciprocally on flexor and extensor motoneurones, either by co-exciting (flexor) or by co-inhibiting (extensor) α- and γ-motoneurones (*Hongo* et al. 1969; *Appelberg* et al. 1975, 1982b).

Fourth, stimulation of the *motor cortex* in the baboon may excite both α- and γ-motoneurones of finger flexor muscles monosynaptically (*Grigg* and *Preston* 1971; *Clough* et al. 1971). Using the indirect method of spindle afferent recording, *Koeze* (1973b) confirmed that skeletomotor and fusimotor neurones to hindlimb flexors of the baboon indeed were frequently co-excited. Interestingly, he also found that in this preparation the motor cortex operated static fusimotor neurones apparently selectively.

Parallel *monosynaptic excitatory projections* could, no doubt, provide a route for tightly coupled coactivation of α- and γ-motoneurones. However, two reservations concerning this interpretation have to be borne in mind. First, tightly coupled and synergistic short-latency effects could easily be overridden by more powerful antagonistic effects of longer latency. Second, even if monosynaptic projections were dominant, they might exist as two functionally separate groups of descending axons, one to the α- and one to the γ-motoneurones. Electrical stimulation in laboratory preparations might, however, lack the resolution to activate these pathways separately; but under physiological conditions intrinsic central circuitry might be capable of doing just that.

Observations at the other end of the spectrum indicate that α- and γ-motoneurones may also be activated independently of each other from central structures: *α-γ independence. Granit* et al. (1955) noted that after ablation or cooling of the anterior lobe of the *cerebellum* α-motoneurones were excited, whereas γ-motoneurones were inhibited (as judged from spindle afferent recordings). A similar decoupling between α- and γ-activity was subsequently found upon electrical stimulation of surface or deep cerebellar structures (*Henatsch* et al. 1964; *Corda* et al. 1966), for instance when γ-fibres were excited while α-fibres were inhibited. *Fidone* and *Preston* (1969) found that on activation of the *corticospinal tract* only two-thirds of the γ-fibres were either co-excited (flexors) or co-inhibited (extensors) with the α-motoneurones supplying the same muscles (see *Agnew* et al. 1963; *Agnew* and *Preston* 1965; *Lundberg* and *Voorhoeve* 1962). The remaining one-third of the γ-fibres showed the reverse pattern of activation compared with the homonymous α-motoneurones.

Additional examples of independent central control of α- and γ-motoneurones will be listed below. Yet with the recognition of static and dynamic γ-motoneurones (*Matthews* 1962), the issue had become more complex (see *Matthews* 1964, 1972). The relevant question was no longer 'α-γ coactivation, yes or no?', but instead 'coactivation, α-γ_S and/or α-γ_D? If so, to what degree?'.

4.2.2 Static and Dynamic γ-Motoneurones

If static and dynamic fusimotor neurones were not independently controlled by descending pathways, they might never have been discovered. The essential observation of *Jansen* and *Matthews* (1962) was that in the decerebrate cat the dynamic and static sensitivity of primary afferents showed sizable and mutually independent fluctuations. These observations clearly demonstrated the potential of central structures to independently control static and dynamic *sensitivity* (*Jansen* and *Matthews* 1962). *Matthews* (1962) then went on to search for, and indeed define, static and dynamic *γ-motoneurones.* The earlier observations could then be reinterpreted as evidence for *selective central control* of these two types of fusimotor neurone.

This concept has since been amply confirmed. In a considerable number of studies, either predominantly static or predominantly dynamic effects were elicited from various parts of the brain, often depending on the state of preparation, the pattern of stimulation or the level of anaesthesia. Common to all these studies was that fusimotor actions were provoked by electrical stimulation of central structures and that they were assessed from afferent rather than γ-efferent recordings (see Sect. 4.1.4).

Stimulation in *MesADC* activated *dynamic* fusimotor neurones apparently selectively. Neither static fusimotor neurones nor – at the relatively low stimulation intensities used – α-motoneurones were operated from this mesencephalic region (see Sect. 4.1.3.1 and *Appelberg* 1981, review).

The *cerebellum* appears to provide tonic and predominantly *static* fusimotor drive (*Gilman* 1969; *Kornhauser* et al. 1982). Two separate output routes seem to be involved, one from the *fastigial nucleus* most likely via the *vestibulospinal tract* (*Kornhauser* et al. 1982; cf. Sect. 4.2.1), the other from the *lateral* and *interposed nuclei,* perhaps via the thalamus (ventrolateral nucleus) and the corticospinal tract (*Gilman* 1969; *Gilman* et al. 1971a,b).

Likewise, GABA-ergic neurones of the reticular zone of the *substantia nigra* appear to contribute mainly *static* fusimotor drive (*Wand* et al. 1981).

From the *caudate nucleus* predominantly *static* fusimotor effects were elicited in an extensor muscle (soleus) (*Vedel* and *Paillard* 1965; *Vedel* 1975). In contrast, in a flexor muscle (tibialis anterior) stimulation of the same nucleus provoked *dynamic* fusimotor effects, but the selectivity of this effect was minimal (*Vedel* and *Coulmance* 1975).

From some regions of the pontine *reticular formation, static* fusimotor effects – and from other regions, *dynamic* fusimotor effects – were elicited. The latter were also obtained on stimulation of the bulbar reticular formation (*Vedel* and *Mouillac-Baudevin* 1969b). These effects were often variable, yet of particular interest for the present consideration is that they could be switched from static to dynamic either on increasing the depth of anaesthesia or simply on repeating the same stimulation procedure (*Vedel* and *Mouillac-Baudevin* 1969a).

The findings of *Yokota* and *Voorhoeve* (1969) suggested that in the cat both *static* and *dynamic* fusimotor neurones could be activated via the *corticospinal tract.* In their hands the balance between the two types of fusimotor effect was invariant. However, *Vedel* and *Mouillac-Baudevin* (1970) found predominantly dynamic fusimotor effects (on stimulation of the motor cortex or the pyramidal tract). Interestingly, for a fraction of the afferents tested the fusimotor action provoked from the motor cortex could switch from dynamic to static. This was true for flexor spindles and, remarkably, also extensor spindles (cf. Sect. 4.2.1, and *Fidone* and *Preston* 1969). These findings betoken again that static and dynamic fusimotor neurones can be activated independently of each other, in this case from the motor cortex.

In conclusion, a number of investigations have provided convincing evidence for the existence of (at least partly) separate descending control of the two types of fusimotor neurones. In most cases the precise pathways involved and the details of the synaptic connections and overall delays remain to be elucidated. Also, given that the data reviewed were obtained in reduced preparations, it cannot be decided to what extent the existing potential of independent control is indeed used during natural motor performance.

4.3 Recurrent Inhibition of γ-Motoneurones

Activity in α-motoneurones is self-limiting, since via their recurrent collaterals they may activate Renshaw cells, which in turn provoke powerful and long-lasting inhibition in those α-motoneurones from which they were excited (*Renshaw* 1941, 1946; *Eccles* et al. 1954; for reviews see *Wilson* 1966; *Hultborn* 1976; *Baldissera* et al. 1981). If Renshaw cells also acted on γ-cells, they might provide an effective co-inhibitory link between α- and γ-motoneurones.

Early attempts to demonstrate recurrent inhibition in γ-motoneurones were not successful (*Granit* et al. 1957; *Hunt* and *Paintal* 1958; *Eccles* et al. 1960; *Voorhoeve* and *Van Kanten* 1962). At the time this apparent lack of recurrent inhibition and the absence of monosynaptic Ia excitation were regarded as clear examples of all-or-none differences between γ- and α-motoneurones. However, only the latter feature has stood the test of time (see Sect. 4.4.2), and what was an absolute distinction has become one of degree, since the occurrence of recurrent inhibition of γ-motoneurones has now been firmly established (see below). Such late recognition is partly attributable to technical difficulties (see Sect. 4.1.1). Yet it also reflects the fact that recurrent inhibition of γ-cells is less frequent and less effective than is the case with α-motoneurones (*Ellaway* and *Murphy* 1981; *Baldissera* et al. 1981).

With both pools of motoneurones (α and γ) being potential sources and recipients, there are theoretically four possible combinations of recurrent inhibitory interactions. In fact, evidence has now accumulated that all four combinations may indeed be operative, although at varying frequency and strength.

α-Onto-α Renshaw inhibition is widespread and powerful. It is most pronounced from α-motoneurones of homonymous and synergistic muscles but also present from some remote muscles, whilst entirely absent from directly antagonistic muscles. Large motoneurones are the main source of recurrent collaterals and thus of Renshaw inhibition, whilst small motoneurones are most effectively inhibited. This is apparently not due to a preferential projection of Renshaw cells to the smaller motoneurones but rather to their greater susceptibility to a synaptic input of a given strength (for further details see *Baldissera* et al. 1981; *Henneman* and *Mendell* 1981; *Burke RE* 1981).

α-Onto-γ recurrent inhibition has now been unequivocally demonstrated (*Brown* et al. 1968a,b; *Ellaway* 1968, 1971; *Grillner* 1969a; *Noth* 1971; *Fromm* et al. 1974; *Fromm* and *Noth* 1976; *Ellaway* and *Murphy* 1981; *Appelberg* et al. 1983d). Yet from all these studies it became quite evident that recurrent inhibition of γ-motoneurones is less frequent and less sizable than that of α-motoneurones. This is surprising, since γ-cells are smaller than even the smallest α-cells, and since they should therefore be much more susceptible than α-cells to inhibitory inputs from Renshaw cells. This would hold up only if Renshaw cells projected to both pools of motoneurones without preference. The relative scarcity of recurrent inhibition in γ-motoneurones therefore indicates that Renshaw cell output is preferentially directed towards α-motoneurones. Nevertheless, when it occurs, recurrent inhibition of γ-cells may also be elicited from other than homonymous sources (*Ellaway* 1971, 1976; *Appelberg* et al. 1983d) but not from antagonist muscle nerves (*Noth* 1971, *Appelberg* et al. 1983d).

γ-Onto-α Renshaw inhibition was until recently believed to be non-existent, since on supramaximal antidromic stimulation of α-axons the inhibitory effects did not further increase (see *Baldissera* et al. 1981 for references). But this was perhaps attributable to saturation in Renshaw cells rather than to a genuine absence of γ-action. Nevertheless, in histological studies no, or only a few, recurrent collaterals from γ-fibres were found (*Cullheim* and *Ulfhake* 1979; *Westbury* 1980, 1982); despite this a weak contribution from γ-cells to the recurrent inhibitory mechanisms could not entirely be discarded. In fact, such weak γ-contribution has indeed been observed by *Kato* and *Fukushima* (1974) when, during a selective block of α-axons, antidromic stimulation provoked modest and infrequent excitation of the Renshaw cells in the appropriate motor nucleus.

γ-Onto-γ recurrent inhibition was also believed to be only a theoretical possibility for the same reasons. However, a few cases of recurrent inhibition of γ-cells, elicited by autogenetic stimulation in the γ- but not in the α-range, have now also been reported (*Appelberg* et al. 1983d).

Thus, recurrent inhibition is clearly most frequently elicited by activity in the large α-motoneurones and least frequently by γ-activity. The most responsive recipients are the small α-motoneurones, whereas the even smaller γ-motoneurones receive surprisingly little Renshaw inhibition.

Both *static and dynamic* γ-motoneurones can be subjected to the recurrent inhibitory control. This was concluded from studies on γ-motoneurones, which were indirectly classified either with the aid of DOPA (*Ellaway* 1971) or of MesADC stimulation (*Appelberg* et al. 1983d). However, the limited size of the samples of γ-cells so investigated does not as yet permit final conclusions on any preferential action of Renshaw cells on one or other type of fusimotor neurone.

The functional significance of Renshaw cells still is a matter of conjecture (see *Baldissera* et al. 1981). Renshaw cells are subjected to widespread excitatory and inhibitory control from virtually all structures involved in motor control (see general reviews, above). It has therefore been proposed that by virtue of being indirectly activated by the Ia fibres they might act as a general gain-regulating system for homonymous and reciprocal Ia reflex actions on α-motoneurones (*Hultborn* et al. 1979), for Renshaw cells are not restricted to inhibiting α- and γ-motoneurones. Instead, they also act on other Renshaw cells as well as on the Ia inhibitory interneurone (see general reviews, above).

Returning to the concept of α-γ coactivation, Renshaw cells may now be seen as contributing a co-inhibitory link – albeit a rather weak one – between α- and γ-motoneurones, whose effectiveness may, however, be modulated from higher motor centres.

4.4 Reflex Control of γ-Motoneurones

The control of fusimotor neurones by peripheral reflex mechanisms bears on questions of general interest. Reflex and descending pathways converging on α-motoneurones may interact in a functionally meaningful way at the level of spinal interneurones (see, e.g. *Lundberg* 1979a,b; *Jankowska* 1979; *Baldissera* et al. 1981). Similar interactions between peripheral sensory feedback and descending commands are likely to shape the final fusimotor outflow to the muscle spindles. Descending rule may then bring functional meaning into what appears to be a large and heterogeneous mosaic of peripheral reflex effects.

Since fusimotor fibres excite spindle afferents, an excitatory pathway from spindle afferents onto their own fusimotor neurones would form a potentially unstable loop, powered by positive feedback (cf. *Houk* 1972). The risk of such loops ever being critically activated can best be assessed on the basis of detailed knowledge of the type and strength of each particular reflex input to the pool of fusimotor neurones, and on the basis of simultaneous recordings of spindle afferent and α- and γ-efferent activity in closed reflex loops (see *Bessou* et al. 1981, 1984).

The concept of α-γ coactivation derived its initial support from the observation of a close similarity in the reflex responses of α- and γ-motoneurones (*Hunt* 1951; *Kobayashi* et al. 1952; *Eldred* and *Hagbarth* 1954; *Granit* 1955). Yet this was before the recognition of static and dynamic γ-cells. The question then was whether reflex coactivation was the guiding rule for either type of fusimotor neurone if this view held up at all.

4.4.1 Size and Composition of Reflexogenic Territories

The reflex control of α- and γ-motoneurones is not identical in all respects. Three main differences may be pointed out before considering the details. First, in the domain of preferred α-responsiveness (muscle group I inputs), γ-motoneurones are rather irresponsive (see Sect. 4.4.2). Second, γ-motoneurones appear to possess wider and more heterogeneous reflexogenic territories, so that it is often difficult to ascribe a meaningful and specific function to individual reflex effects (see below). Third, γ-motoneurones often possess lower reflex thresholds than α-motoneurones, so that weak stimulation may activate γ-fibres selectively. Moreover, in reduced preparations γ-fibres may be spontaneously active, when α-fibres are not.

Similarities. The earliest studies of reflex responses in γ-fibres emphasized their similarity with the familiar reflexes as seen in α-motoneurones (see *Hunt* and *Perl* 1960, review), especially those which are elicited from the *skin* or *subcutaneous structures.*

Touching or squeezing of the ipsilateral footpad provoked *flexor reflex* responses in both α- and γ-motoneurones (co-exciting flexors and co-inhibiting extensors). From the contralateral side the same stimuli gave rise to *crossed extensor reflexes,* co-exciting extensor and co-inhibiting flexor α- and γ-motoneurones (*Hunt* 1951; *Kobayashi* et al. 1952; *Hunt* and *Paintal* 1958). Likewise in the *extensor thrust* reaction (*Hunt* 1951), the two types of motoneurone were co-excited. Similarly, in spatially *focused cutaneous reflexes* (*Hagbarth* 1952), α- and γ-motoneurones were co-excited from the skin region above the target muscle and co-inhibited from elsewhere, especially from the skin areas associated with the antagonist muscles (*Eldred* and *Hagbarth* 1954; *Hunt* and *Paintal* 1958). In all these examples γ- but not α-fibres tended to be spontaneously active, and these γ-fibres were more readily activated than the α-fibres to the extent that selective γ-responses were often encountered.

Differences. Some of the differences in reflex behaviour of α- and γ-motoneurones are qualitative; others may be only quantitative and attributable to differences in reflex threshold (but see Sect. 4.2.1).

Lower γ-thresholds have been encountered for a wide range of reflexes (e.g. from skin, above; pinna, below; muscle, *Wuerker* and *Henneman* 1963; *Post* et al. 1980; *Bessou* et al. 1981). In the extreme, γ-responsiveness may approach saturation when α-motoneurones only start being recruited (*Post* et al. 1980).

From the outset it was clear that the myotatic stretch reflex had no counterpart in the γ-domain. Instead *Hunt* (1951) found that in the decerebrate cat γ-motoneurones were inhibited by stretch of their target muscle: *autogenetic inhibition* (see Sect. 4.4.2).

The flexor/crossed extensor reflex pattern was not found in all muscles. In *extensor digitorum longus* (a functional flexor) γ-motoneurones, but not α-motoneurones, behaved true to their name instead of their function (*Hunt* and *Paintal* 1958).

In the spinal cat, lower reflex thresholds were only present in those γ-fibres that exhibited *spontaneous activity.* In contrast, the γ-fibres, which like the α-fibres were silent at rest, had even higher reflex thresholds than the α-motoneurones (*Hunt* and *Paintal* 1958).

The *pinna reflex* (*Granit* et al. 1952), provoked by twisting the ear, may vigorously excite fusimotor neurones without any (*Andrew* et al. 1979), or with only spurious (*Granit* et al. 1952), coactivation of α-motoneurones.

The *reflexogenic territories* of γ-motoneurones appear to be larger and more heterogeneous than those of α-motoneurones. A few examples are listed below and more could easily be added (without adding insight, though). However, what bears emphasis is that such diverse inputs are not normally seen to excite α-motoneurones, and that in most cases it is difficult to ascribe a specific function to each particular observation.

Pulling on *scalp muscles* or on the *eye bulb* (during surgery) may elicit striking spindle afferent responses from hindlimb muscles (*Granit* et al. 1952). *Distension* of the *bladder* has also been reported to excite γ-fibres, not only those of flexors (*Abdullah* and *Eldred* 1959) but also those of extensors (*Grillner* 1969a), whereas *Evans* (1963) found both excitatory and inhibitory responses of γ-fibres belonging to either one or the other type of muscle. Moreover, manipulation of *trunk muscles*, pinching of the far end of the *tail*, poking of the *intestines* with a rectal thermometer, or manipulations of the *periosteum* of the iliac crest or exposed vertebrae have all been seen to excite hindlimb extensor γ-motoneurones (*Grillner* 1969a). Likewise, when *tendons* are squeezed, γ-motoneurones may be excited, probably from nociceptive afferents (*Hunt* and *Paintal* 1958; *Fromm* et al. 1974). Further, in spinal cats reflex responses of γ-efferents to local changes in *temperature*, either in the skin or in the spinal cord, have also been reported (*Sato* and *Hasegawa* 1977; *Sato* 1981). Mainly dynamic fusimotor efferents seem to be activated by maintained changes in temperature – both increases and decreases (*Schäfer* and *Schäfer* 1973; *Sato* 1983) – whereas static fusimotor neurones appear to be inhibited during cold shivering (*Schäfer* and *Schäfer* 1973). Finally, excitation of joint afferents has been reported to activate ipsilateral static (*McIntyre* et al. 1978) and contralateral dynamic (*Appelberg* et al. 1979) fusimotor neurones.

Grillner (1969a,b) pointed out that the inputs from such wide and heterogeneous reflexogenic territories might merely provide a tonic background drive to the excitatory and inhibitory interneurones acting on γ-motoneurones. These would then be the substrates through which selective descending pathways might impose their orderly rule. However, much as it is attractive, this proposal is at variance with the earlier findings of *Shimazu* et al. (1962a,b) and of *Appelberg* and *Kozary* (1963) that various descending systems exerted diffuse and non-reciprocal facilitatory effects on both flexor and extensor γ-motoneurones. *Grillner's* hypothesis would therefore greatly gain if examples of the requisite selective reflex suppression or facilitation could be demonstrated on stimulation of individual descending pathways.

The apparently much larger reflexogenic territories of γ- than α-fibres could simply be due to a consistently lower reflex threshold of γ-motoneurones. Comparisons with published data on skeletomotor reflexes are of little help, in view of the familiar dependence of reflex responsiveness on the level of anaesthesia and on the state of the preparation. Only systematic comparisons of reflex responses in pairs or pools of a α- and γ-fibres (see, e.g. *Hunt* 1951; *Eldred* and *Hagbarth* 1954; *Bessou* et al. 1981, 1984) will eventually settle this issue. Whatever the outcome, this cannot distract from the fact that there are important differences in the reflex patterns of static and dynamic γ-motoneurones.

4.4.2 Reflex Actions from Muscle Group I Afferents

In his pioneering study, *Hunt* (1951) described autogenetic inhibition of γ-motoneurones in the decerebrate cat, which was elicited by muscle stretch. At the time it was not known which receptor afferents were responsible for this inhibitory effect, but the finding was important, since it was in striking contrast with the autogenetic excitation of α-motoneurones by spindle primary afferents. However, on reinvestigating the phenomenon in spinal preparations *Hunt* and *Paintal* (1958) failed to confirm the earlier observation when they studied otherwise highly responsive γ-motoneurones. The important new finding of this later study was that not one of the – non-autogenetic – excitatory effects encountered was monosynaptic.

Intriguing as they were, these findings have since been confirmed. *Absence of monosynaptic excitation* was also reported by *Eccles* et al. (1960), by *Grillner* et al. (1969) and – with one or two borderline exceptions – by *Appelberg* et al. (1983a). Thus, in this respect the reflex organization of γ-motoneurones is clearly different from that of α-motoneurones, where monosynaptic Ia excitation is pronounced (*Eccles* et al. 1957a,b; *Brown* and *Fyffe* 1981; *Burke RE* 1981). This distinction is all-or-none rather than merely of degree.

The *occurrence of autogenetic inhibition* has since also been widely confirmed. With natural stimulation, it was found on stretching the muscle (*Fromm* et al. 1974; *Fromm* and *Noth* 1976; *Ellaway* and *Trott* 1978), on vibrating the muscle (*Brown* et al. 1968a; *Fromm* and *Noth* 1976; *Ellaway* and *Trott* 1978), or on eliciting twitches in the target muscle of the γ-fibres (*Ellaway* and *Trott* 1976; *Ellaway* et al. 1979; *Ellaway* and *Murphy* 1980). These findings were further supported by studies using graded electrical stimulation of the autogenetic muscle nerve within the range of muscle group I intensities (*Grillner* et al. 1969; *Appelberg* et al. 1983a; *Noth* 1983). However, none of these investigations permitted final conclusions concerning the types of afferents which were responsible for the autogenetic inhibition. The most likely possibility seems to be that both *Ib afferents* from tendon organs (activated by carefully adjusted muscle twitches, above) and *Ia primary spindle afferents* (activated by small-amplitude vibration) are contributing. The Ia afferents might act via excitation of α-motoneurones and subsequent activation of Renshaw cells, since vibration-induced and antidromic inhibition were highly correlated (*Fromm* and *Noth* 1976; see also *Noth* 1983).

Fromm and *Noth* (1976) plausibly explained the paradoxical findings of *Hunt* and *Paintal* (1958) by demonstrating that in the spinal state the inhibitory effects were merely masked. For after the administration of DOPA (sensitizing both static and dynamic γ-motoneurones in extensor muscles; *Grillner* 1969a) autogenetic inhibition was readily elicited.

Other muscle group I reflex effects have also been described. *Autogenetic* afferents may also provoke *excitation* (*Fromm* and *Noth* 1976; *Trott* 1976; *Ellaway* and *Trott* 1976, 1978; *Appelberg* et al. 1983a). Moreover, group I *inhibition* has been encountered from a wide range of heteronymous nerves (*Proske* and *Lewis* 1972; *Fromm* et al. 1976; *Ellaway* and *Murphy* 1980; *Appelberg* et al. 1983a).

However, the most important aspect of *group I reflex action* on γ-motoneurones is that it is very *rare*, compared with the much higher incidence of reflex actions from cutaneous and joint receptors as well as muscle group II and III afferents (*Appelberg* et al. 1977, 1983a,b,c; *Noth* 1981). Moreover, muscle group I reflex action was equally ineffective for static and dynamic γ-cells, as classified from MesADC (see Sect. 4.1.3.1; *Appelberg* et al. 1983a).

In conclusion, γ-motoneurones differ from α-motoneurones in their reflex excitability by muscle group I afferent fibres (spindle primary and tendon organ afferents): first, by the absence of monosynaptic connections and second, by their generally poor responsiveness.

4.4.3 Reflex Actions from Muscle Group II Afferents

For a long time the sole known role for secondary muscle spindle afferents was their contribution to flexor reflex responses. As part of the flexor reflex afferent (FRA) system (comprising high-threshold muscle, skin, and joint afferents) they were thought to mainly provoke excitation of flexor, and inhibition of extensor, α-motoneurones, irrespective of whether their parent muscles were flexors or extensors (*Eccles* and *Lundberg* 1959; *Holmqvist* and *Lundberg* 1961). However, the generality of this view was challenged when *Matthews* (1969) suggested, on indirect evidence, that secondary spindle afferents contributed to the stretch-evoked excitation of α-motoneurones. In due course this was indeed confirmed with the aid of the spike-triggered average technique (*Kirkwood* and *Sears* 1974, 1975; *Stauffer* et al. 1976).

Recent evidence from studies on the reflex action of muscle group II afferents on γ-motoneurones suggests that these are not organized according to the dominant FRA- or α-pattern (*Appelberg* et al. 1983b,c). First, γ-motoneurones were more frequently excited than inhibited on electrical stimulation of muscle group II afferents (see Fig. 13). This was true not only for those supplying flexor muscles (conforming with the FRA pattern) but also for those to extensor muscles (at variance with the FRA pattern). Second, muscle group II action on dynamic γ-cells was clearly more potent than that on static γ-cells. This was again found for both flexor and extensor γ-motoneurones (*Appelberg* et al. 1977, 1983b).

These results have been confirmed along separate lines. *Noth* and *Thilmann* (1980) and *Noth* (1981), also using electrical stimulation within the appropriate range of intensities, agreed that both autogenetic and heteronymous (from PBSt, posterior biceps semitendinosus) muscle group II afferents were rather effective in activating non-classified γ-fibres supplying an extensor muscle (triceps surae). Further, *Appelberg* et al. (1982a) showed that natural stimulation (muscle stretch) indeed activated mainly dynamic fusimotor neurones. In view of the scarcity of group I reflex action this effect was attributable to group II afferents.

These findings strongly suggest that γ-reflexes are not organized according to the dominant FRA or α-pattern as was suggested by *Grillner* et al. (1969) (on limited number of observations). Moreover, these data provide another important example to support the view that static and dynamic fusimotor neurones may be controlled by separate central mechanisms.

4.4.4 Reflex Actions from Muscle Group III Afferents

The fine, high-threshold muscle afferents which conduct at velocities in the group III (myelinated fibre) and group IV (unmyelinated fibre) range comprise chemoreceptor, nociceptor, and stretch receptor afferents (*Paintal* 1960; *Bessou* and *Laporte* 1961; *Iggo* 1961; *Mense* 1977; *Kniffki* et al. 1978; *Ellaway* et al. 1982). These fibres subserve a wide range of circulatory and respiratory functions (see *Kniffki* et al. 1978). In electrophysiological terms the group III and IV afferents belong to the FRA system, whose function may extend beyond the mere generation of flexor reflexes. Thus, *Jankowska* et al. (1967a,b) suggested that tonic FRA activity might be the driving force provoking rhythmic alternating motor activity in spinal interneurone networks (half-centre model; see *Lundberg* 1979b). Recent evidence has indeed confirmed that muscle group III and IV afferents, activated by adequate stimuli, exert a sizable effect on fictive locomotion in spinal cats (*Kniffki* et al. 1981).

Seen against this background and given the sheer number of these high-threshold afferents (*Boyd* and *Davey* 1968; *Richmond* et al. 1976; *Schmidt* et al. 1981), the question whether γ-motoneurones were also influenced by muscle group III afferents is evidently relevant. *Voorhoeve* and *Van Kanten* (1962) and *Grillner* et al. (1969) concluded from scarce evidence that group III reflex actions (much as group II effects) on extensor γ-motoneurones followed the FRA pattern. However, the detailed reinvestigation of this matter by *Appelberg* et al. (1977, 1983c) failed to support this view.

The important new finding was that in contrast with the group II responses group III reflex action was more frequent in static than in dynamic γ-cells. Moreover, as was the case for group II actions, excitation prevailed over inhibition, not only for flexor but also for extensor γ-motoneurones. This was in agreement with the study of *Ellaway* et al. (1982), who also encountered predominantly excitatory reflex actions of muscle group III afferents on non-classified γ-fibres. In keeping with earlier reports (*Bessou* and *Laporte* 1961; *Mense* and *Stahnke* 1978, see also 1983), *Ellaway* et al. (1982) further emphasized that the adequate stimulus for group III afferents could be low-threshold mechanical events, such as the tension transient of a muscle twitch.

Thus, much in line with the findings from muscle group II reflex effects, the results obtained in the group III range confirm that fusimotor reflexes may differ from those seen in skeletomotor neurones by not conforming with the standard FRA pattern. Moreover, the group III data further support the concept of non-uniform reflex control of static and dynamic fusimotor neurones.

4.5 Conclusions

The findings reviewed in this section have all been obtained in anaesthetized or even further reduced preparations, often with the aid of electrical rather than natural stimulation. Therefore, they may tell us more about the potential for control than about its actual use (see Sect. 5) during natural movements. Nevertheless, the potential laid down in the brain seems sufficiently complex to make it unlikely that the control of α- and γ-motoneurones is always either tightly coupled or entirely independent. Thus the reflex effects of cutaneous afferents have so far shown a relatively high degree of similarity for the two types of motoneurones, whereas the reflexes from muscle afferents show a much higher degree of incongruence, both between α- and γ-motoneurones and between static and dynamic γ-motoneurones.

As regards the risk of *reflex instability* arising from positive feedback from spindle afferents onto their own fusimotor neurones (autogenetic excitation), it bears emphasis that all the potentially unstable loops are poorly developed (Ia onto γ_D, Ia onto γ_S, and II onto γ_S). The only safe connection is the group II- (spindle secondary) onto-γ_S projection, since it does not form a closed loop. Interestingly, this is clearly the most effective among the four possible reflex connections. Simultaneous recordings of spindle afferent and α- and γ-efferent discharge during imposed movements in decerebrate preparations (*Bessou* et al. 1984) indeed support the view that the risk of reflex instability is small.

At the level of the spinal cord there are considerable *differences* in the reflex control of *static* and *dynamic* fusimotor neurones. Similar differences are also present in the control by descending pathways, where the evidence available indicates that static fusimotor neurones are more closely coupled with α-motoneurones than dynamic fusimotor neurones are. It may be concluded that the central nervous system does indeed possess the means to control the activity in fusimotor neurones independently from that in skeletomotor neurones. The independence is not absolute but is of degree, yet the degree of flexibility is greater for dynamic than for static fusimotor fibres.

Figure 13 gives a schematic overview of the main features that were emphasized in this section (see legend). Its purpose is to draw attention to the *complexity of interactions* which take place at two separate levels

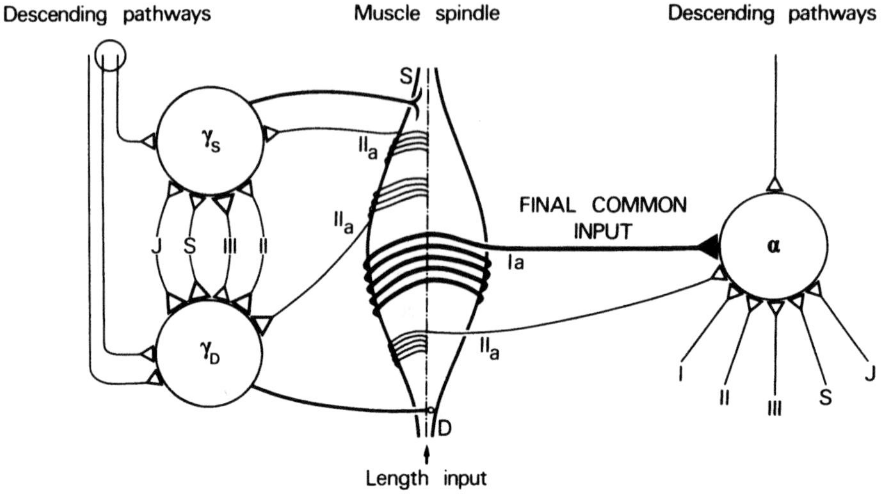

Fig. 13. The complexity of interactions at the level of the γ-motoneurones and of the muscle spindle. On the *left,* the scheme shows the multisensory convergence on static (γ_S) and dynamic (γ_D) γ-motoneurones, from autogenetic (*IIa*) and heteronymous (*II*) muscle group II, from muscle group III (*III*) as well as from joint (*J*) and skin (*S*) afferents. Descending pathways which act in parallel on both types of fusimotor neurone are *encircled.* The size of the schematic synaptic boutons indicates the relative size of the reflex action.

At the level of the spindle (*middle*), the polymodal signals from the γ-motoneurones interact with the mechanical input (muscle length). The primary (*Ia*) and secondary (*IIa*) spindle afferents then send a complex message to the central nervous system and to the α-motoneurones (*right*). The secondary afferents further constitute a link between the two types of γ-cells by being acted upon from static efferents and by acting on dynamic γ-motoneurones. The weak autogenetic actions from group I afferents onto γ-cells are not illustrated. Each input category is represented by a single channel which may stand for a number of parallel pathways and for both excitatory and inhibitory connections. (From *Appelberg* et al. 1983c, with permission)

of the γ-loop, first, in the γ-motoneurone itself and second, in the muscle spindle.

γ-Motoneurones receive synaptic inputs from widely ranging sources, including a number of descending pathways (see Sect. 4.2) and numerous reflex inputs (Sect. 4.4.1). During natural movements at least some of these input pathways are bound to be activated. Whilst nothing is known about any movement-related activity in the particular descending tract neurones to γ-motoneurones, the activity of some of the reflexogenic peripheral inputs is bound to be movement related. This has been amply demonstrated for the afferents from moved muscles (see Sect. 5). Recently cutaneous afferents have also been shown to respond to natural movements (*Hulliger* et al. 1979; *Appenteng* et al. 1982a) and likewise joint receptor afferents should respond, at least to extreme range movements, as judged by their general characteristics (*Burgess* and *Clark* 1969; *Grigg* 1975; *Grigg* and *Greenspan* 1977). These reflex inputs are bound to interact with descending inputs, so that the outflow from fusimotor neurones during movement will hardly reflect descending commands alone. Instead, the fusimotor activity is likely to be a highly processed signal with polymodal characteristics (*Johansson* 1981; *Appelberg* et al. 1983c).

In conclusion, spindle endings are not simple stretch receptors for two reasons. First, in addition to stretch, they are also excited by fusimotor drive. Second, this driving signal on its own is likely to have polymodal properties, since it may be shaped both by descending commands and widely ranging peripheral reflex actions.

5 Spindle Afferent and Fusimotor Activity During Natural Movement

Experiments on reduced preparations are unlikely to provide final answers for the understanding of complex neural systems. For instance, anaesthetics may induce profound changes in function, even at the level of simple reflexes. Any theory on motor control therefore has to be put to a final test in intact and unrestrained organisms. In the following, the consideration of spindle behaviour during natural movement is largely restricted to chronic recordings obtained in freely behaving animals (Sect. 5.2) and to microneurographic recordings during truly voluntary movements in man (Sect. 5.3). The obvious merits of such recordings are, however, somewhat qualified by the methodological problems encountered (for critical review see *Prochazka* and *Hulliger* 1983). Amongst those, the most serious limitation is that direct recordings from adequately identified and classified (static or dynamic) fusimotor neurones have not as yet been achieved (owing to technical constraints, see references above). In contrast, in reduced preparations recordings from γ-motoneurones have been feasible. These studies have sufficiently influenced current thinking to justify a brief summary in this context (for more detailed reviews, see *Sears* 1973; *Prochazka* 1981).

5.1 Stereotyped Movements in Reduced Preparations

Certain stereotyped rhythmic movements, such as chewing or licking, stepping and breathing, have been intensely studied in reduced (i.e. anaesthetized, decerebrate or spinal) preparations. Common to all these studies is the finding that γ-activity could be closely coupled with α-activity. However, in addition to such rhythmic coactivation, γ-activity could also be tonically sustained.

Respiration. Rhythmic and tight coactivation of α- and γ-fibres has been observed for both inspiratory and expiratory muscles (*Eklund* et al. 1964; *Sears* 1963, 1964; *Corda* et al. 1966), yet γ-activity could also be maintained at a steady level, whilst α-activity was rhythmically modulated. These patterns were not necessarily fixed properties of distinct classes of γ-motoneurones, since on several occasions individual γ-fibres were seen to change from one form of activity to the other (*Eklund* et al. 1964).

Moreover, recordings from spindle afferents clearly showed that rhythmic coactivation of fusimotor fibres could be sufficiently widespread and powerful to offset spindle unloading during active muscle shortening (*Critchlow* and *Von Euler* 1963), although such compensatory action was not equally pronounced in all muscles (*Corda* et al. 1965).

These findings then suggested that in the case of respiratory movements closely linked descending activation of α- and γ-motoneurones played an important part, but the presence of tonic γ-firing during rhythmic skeletomotor activity indicated that such tight coupling was not the only choice.

Jaw Movements. Spindle afferent responses during induced cyclic jaw movements in anaesthetized cats could either show signs of modulated fusimotor outflow, strong enough to compensate for unloading during shortening, or they could show stretch-sensitive behaviour, compatible with largely tonic fusimotor drive (*Davey* and *Taylor* 1966; *Appenteng* et al. 1980). Recordings from γ-fibres revealed two main patterns, 'modulated' and 'sustained', which were tentatively ascribed to static and dynamic γ-motoneurones (*Appenteng* et al. 1980; *Gottlieb* and *Taylor* 1983). On this view, modulated static fusimotor outflow (perhaps a 'temporal template of the intended movement') was responsible for maintenance of spindle firing during shortening, whereas tonic dynamic drive was set to provide sensitive monitoring of incremental length changes (*Taylor* and *Appenteng* 1981; *Gottlieb* and *Taylor* 1983). Thus coactivation was restricted to one class of γ-motoneurones (modulated, static), whilst for a second class (sustained, dynamic) activity during movement was independent of skeletomotor neurones.

Locomotion. Evidence for rhythmic coactivation of skeletomotor and fusimotor neurones has also been obtained from spindle afferent recordings during induced locomotion. This applied for both flexor and extensor muscles and was found in precollicular-decerebrate (*Severin* et al. 1967; *Severin* 1970), decorticate (*Perret* and *Buser* 1972; *Perret* and *Berthoz* 1973), as well as spinal (*Sjöström* and *Zangger* 1975, 1976) preparations. Direct recordings from efferent fibres indeed confirmed a closely parallel pattern of α- and γ-discharge (*Sjöström* and *Zangger* 1975, 1976) where during each step, in keeping with the notion of a lower γ-threshold, γ-activity preceded and outlasted α-activity. Thus, in reduced locomotor preparations α-γ coactivation appears to be the rule, yet this rule does not seem to hold up in freely moving animals (see Sect. 5.2.5).

5.2 Recordings from Chronically Implanted Cats and Monkeys

5.2.1 Techniques and Unit Identification

Mainly two techniques have been employed for chronic recordings from conscious cats and monkeys, and they are briefly described below (for details see *Prochazka* 1981; *Luschei* and *Goldberg* 1981; *Prochazka* and *Hulliger* 1983).

5.2.1.1 Midbrain Recordings. Compared with limb nerves, the trigeminal (Vth cranial) nerve is unique in that the cell bodies of afferent neurones from jaw-closing muscles are assembled in a circumscribed midbrain area, the mesencephalic nucleus (Mes V) rather than in the gasserian ganglion (*May* and *Horsley* 1910; *Thelander* 1924; *Corbin* 1940; *Szentagothai* 1948). Since this nucleus is readily accessible, it has been used for recordings of muscle afferent activity during normal biting and chewing movements, using chronically implanted microelectrodes (*Matsunami* and *Kubota* 1972; *Taylor* and *Cody* 1974; *Goodwin* and *Luschei* 1975; *Cody* et al. 1975; see also *Taylor* and *Davey* 1968, technique).

In the *cat* the trigeminal midbrain nucleus contains cell bodies of spindle but not of tendon organ afferents (*Jerge* 1963; *Cody* et al. 1972). It is therefore unlikely that during chronic recordings tendon organ afferents are mistaken for spindle afferents. However, the distinction between primary and secondary spindle afferents has been more problematic. Reliable measurements of conduction velocity are difficult (short conduction distance; *Cody* et al. 1975; but see *Lund* et al. 1979). Moreover, the afferent fibre diameter spectrum lacks bimodality, so that conduction velocity estimates are of limited usefulness for differentiation (*Smith* et al. 1968; *Inoue* et al. 1981; *Morimoto* et al. 1982). A subdivision of spindle afferents can therefore only be tentative. Classification of spindle afferents with the aid of adequate ramp stretches should be feasible, theoretically, but the limited stability of the recordings may not permit such tests on a routine basis.

In the *monkey*, direct anatomical evidence for the absence of tendon organ afferents from Mes V is not as yet available (see *Luschei* and *Goldberg* 1981). Moreover, reliable maximum twitch tests are not readily performed in jaw-closing muscles (*Cody* et al. 1972). The differentiation of spindle from tendon organ afferents has therefore been based on their responses to stretch and palpation (*Matsunami* and *Kubota* 1972; *Goodwin* and *Luschei* 1975) and must be considered somewhat uncertain.

5.2.1.2 Dorsal Root Recordings. The technique of chronic recording from dorsal root afferent fibres or ganglion cells using implanted wire electrodes was introduced independently by *Prochazka* and co-workers in 1975, and by *Loeb* and colleagues in 1976 (see *Prochazka* 1981; *Prochazka* and *Hulliger* 1983, for further details). Initially this method was used in freely moving cats, but recently a modified version using implanted recording chambers and microelectrodes has also been employed in monkeys performing learned motor tasks (*Schieber* and *Thach* 1980). The technique stands out by its remarkable stability, which often has permitted recordings from the same afferent over periods in excess of 24 h. It has therefore been possible to temporarily anaesthetize chronically implanted cats and to perform most of the desirable identification tests. Thus spindle and tendon organ afferents are routinely separated on the basis of their responses to an electrically induced muscle twitch. Moreover, ***Prochazka*** and co-workers have used suxamethonium to further differentiate primary and secondary spindle afferents (***Rack*** and ***Westbury*** 1966; ***Dutia*** 1980). ***Hoffer*** et al. (1981a,b) have recently developed a triple recording technique to estimate axonal conduction velocity for the classification of spindle afferents. This method relies on the averaging of compound potentials recorded from two distal recording sites within a nerve cuff, with the averaging device being triggered by single-unit impulses recorded simultaneously from the dorsal roots.

5.2.2 Spindles as Stretch Receptors During Voluntary Movement?

If the performance of voluntary movement were to be aided by spindle afferent feedback, fusimotor activity would normally have to be rather strongly modulated, so as to counteract spindle unloading during active shortening movements. This could then provide an ongoing working discharge in spindle afferents which might form the basis of reflex- or servo-assisted control of movement (see *Phillips* 1969; *Matthews* 1972; *Stein* 1974). Otherwise, without such modulated and compensatory fusimotor outflow spindle activity would normally be dominated by responses to muscle stretch and release. In this case the main segmental reflex action of spindle afferents, the stretch reflex, would have to be suppressed, since by its homeostatic nature this would tend to suppress rather than assist the execution of movements.

Up until now the main findings from studies on chronically recorded cats and monkeys were that the discharge patterns of spindle afferents during active movements were dominated by their responses to muscle stretching and that the afferents were silenced or only moderately activated during muscle shortening when the parent muscle was actively contracting. In other words, spindle afferents by and large behaved like *stretch receptors.* Moreover, whilst signs of fusimotor activity concurrent with the activation of the parent muscle were often clearly detectable, this did not suffice to offset the unloading effect of the active shortening movement (*Goodwin* and *Luschei* 1975; *Cody* et al. 1975; *Prochazka* et al. 1976, 1977; *Loeb* and *Duysens* 1979, for extensor muscles; *Lewis* et al. 1979a,b; *Prochazka* et al. 1979; *Prochazka* and *Wand* 1981; *Taylor* and *Appenteng* 1981; *Larson* et al. 1983).

Yet this is by no means to say that spindle afferent responses were insignificant. On the contrary, the responses – especially of presumed or identified primary afferents – from jaw-closing muscles during chewing, biting and licking; or from hindlimb muscles during walking, stepping, placing reactions, landing from free fall and paw shake; or from tail muscles during tail movements were impressive by any standard. Thus, peak *discharge rates* at the onset or during muscle stretch of 300 impulses/s were seen with chewing movements in alert monkeys (*Goodwin* and *Luschei* 1975; cf. also *Larson* et al. 1981), and peak rates of between 240

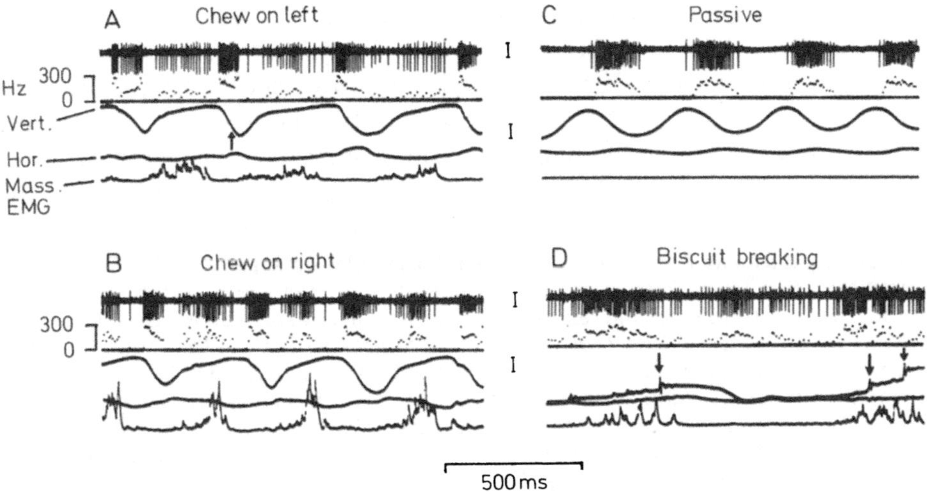

Fig. 14A–D. Responses of a spindle afferent from the right masseter muscle of a monkey during chewing on the left (**A**), on the right (**B**), during passively imposed movements (**C**, light ketamine anaesthesia), and during biscuit breaking (**D**). *Top trace,* unitary discharge; *second trace,* instantaneous discharge rate; *third trace,* vertical jaw movement (shortening plotted upwards); *fourth trace,* horizontal jaw movement (excursion to the right plotted upwards); *bottom trace,* integrated EMG. (From *Goodwin* and *Luschei* 1975, with permission)

and 600 impulses/s during muscle stretching were a distinguishing feature of presumed primary afferents from jaw-closing muscles studied in cats during eating (*Cody* et al. 1975). Similar rates were repeatedly reported for primary afferents from cat hindlimb muscles during various types of spontaneous or imposed and resisted movements (see Figs. 15, 16A) (*Prochazka* et al. 1976, 1977; *Loeb* and *Duysens* 1979; *Prochazka* et al. 1979; *Prochazka* and *Wand* 1981).

A common finding of all chronic recordings in unrestrained animals is that the discharge patterns during spontaneous movements showed a remarkable degree of *variability* (see Fig. 15A, and references above). This was observed both when comparing different afferents (for the same type of movement) and when comparing different movements whilst recording from the same afferent (e.g. *Goodwin* and *Luschei* 1975; *Cody* et al. 1975; *A. Prochazka,* personal communication). Apart from trivial factors to account for this (variability of unit properties or movement profiles) such variety might more fundamentally depend on the *type of movement* performed. Convincing examples have been illustrated, e.g. by *Cody* et al. (1975; eating vs licking the lips, temporalis spindle afferent) and by *Goodwin* and *Luschei* (1975; chewing vs biscuit breaking, masseter afferent, see Fig. 14). In the latter example, the responses were strikingly different between routine chewing (Fig. 14A and B, stretch-dominated response during jaw opening, plotted downwards) and the less stereotyped biscuit breaking (Fig. 14D, 'coactivation' response during muscle shortening, plotted upwards). However, the interpretation of such examples is difficult, since it cannot be decided with certainty whether these differences arose from a fundamental change in fusimotor strategy or merely from an unmasking of fusimotor drive during slower movements, as might be the case with the example of Fig. 14 (see also *Taylor* and *Appenteng* 1981).

Loeb and *Duysens* (1979) and *Loeb* and *Hoffer* (1981) suggested that part of the variety seen in afferent responses was attributable to fundamental differences in fusimotor outflow to flexor and extensor muscles. Their data seemed to support the earlier view of *Perret* and *Berthoz* (1973; cf. also *Cabelguen* 1981) that during locomotion in reduced preparations *flexor* muscle activity was accompanied by powerful and predominantly *static,* and *extensor* activity by predominantly *dynamic* fusimotor action. Given the difference in excitatory strength between the two types of fusimotor neurone (Sect. 3.2.3), this might explain why spindle afferents behaved like stretch receptors in extensor muscles, whilst being more readily excited during shortening in flexor muscles. However, the conclusion by *Loeb* and co-workers is not supported by other studies on flexor muscles (*Lewis* et al. 1979a; *Prochazka* et al. 1979), since in these, primary afferents showed the same, stretch-response dominated behaviour as in extensor muscles. Whatever the truth, differential fusimotor control of flexors and extensors seems not to be a general rule.

Whilst on the whole, spindle afferent discharge during movement was dominated by responses to muscle stretch, this is by no means to deny the likely presence of fusimotor activity. However, if this included any modulated component, then it was in most cases too weak to dominate spindle discharge at the expense of the stretch responses. The evidence to support this view stems from comparisons between active and passively imposed movements (see Fig. 14; also *Goodwin* and *Luschei* 1975; *Cody* et al. 1975; *Loeb* and *Hoffer* 1981; *Prochazka* et al. 1977; *Prochazka* and *Wand* 1981). This conclusion is further borne out by the repeated observation of primary spindle afferents discharging during slow shortening movements, even if at moderate rates, and by the finding of *Hoffer* and *Loeb* (1983) that afferent discharge during shortening could be abolished and that pronounced discharge during stretch could be reduced during a selective block of γ-efferents by lidocaine.

5.2.3 Importance of the Speed of Movement

Observations, like in the example of Fig. 14, led *Prochazka* et al. (1979) to propose that during unobstructed movements the speed of shortening was decisive in determining whether any concurrent fusimotor outflow would exert itself by manifest excitation of the spindle afferents. It appeared that most published data were compatible with the generalization that for shortening movements at below 20% of a muscle's resting length/s fusimotor activity could be clearly detectable, if not dominant. For faster shortening movements, on the other hand, the effects of spindle unloading seemed to prevail, leaving it open whether phasic or tonic fusimotor activity was still present (see *Prochazka* 1981).

This generalization was in line with known spindle properties (see Sects. 3.1.1, 3.1.2, and 3.2.3). For repetitive movements of a given amplitude, spindle afferent responses are known to increase with the frequency and thus with the speed of the movement, whereas the modulation of afferent discharge provoked by modulated fusimotor drive decreases with increasing repeat frequency (*Andersson* et al. 1968b; *Chen* and *Poppele* 1978). Thus, whilst it is known that for slow movements modulated fusimotor drive can offset spindle afferent responses to concomitant variations in muscle length (*Lennerstrand* and *Thoden* 1968c; *Baumann* and *Hulliger*, unpublished), this compensating capacity of fusimotor neurones is bound to reach a limit as the repeat frequency of modulation increases. Yet the question remained whether the observed critical speed of 20% resting length/s corresponded to this inherent limit of fusimotor action. Preliminary evidence indicates that this is not the case, since on activation of the better part of the total fusimotor supply to a spindle at extreme stimulation rates, primary afferent discharge could still be maintained even during muscle shortening at the fastest physiological speeds (*Appenteng* et al. 1982b).

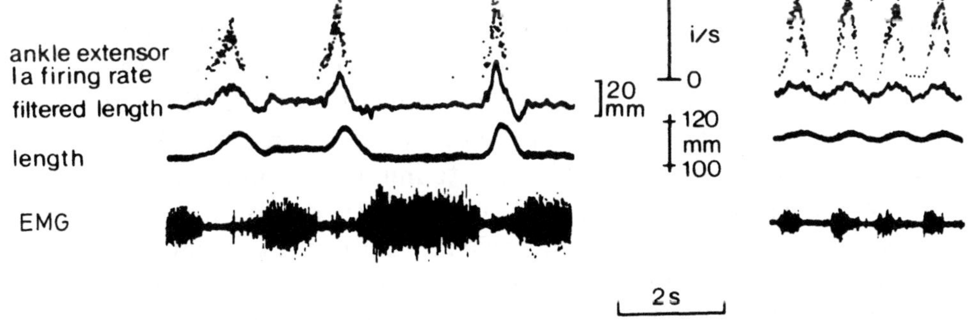

Fig. 15A,B. Ankle extensor primary afferent (**A**) during three voluntary flexion-extension movements (see video tracings, *top*), and during imposed flexion-extension movements (**B**), which were resisted by the animal. Instantaneous discharge rate as in Fig. 14. The filtered length signal was generated with a suitably adjusted high-pass filter in order to mimic the transducing properties of primary spindle afferents (see text). (From *Prochazka* and *Wand* 1981, with permission)

Thus, the observed critical speed of 20% resting length/s reflects choice and not necessity. In other words, during voluntary movements the fusimoto system seems, on present evidence, not to be activated at its full capacity to offset spindle afferent modulation as elicited by muscle stretch and release.

5.2.4 Static and Dynamic Fusimotor Activity During Voluntary Movement

In some clear-cut cases reasonably safe inferences can be made from afferent discharge patterns on the type of fusimotor activity (static/dynamic) which may have accompanied a particular movement. However, more frequently such conclusions stand on uncertain grounds and must remain speculative. *Prochazka* and *Wand* (1981) approached this problem along more quantitative lines using electronically filtered length signals to mimic the transducing properties of spindle afferents and to obtain estimates of dynamic sensitivity of spindle afferents during natural movements. They concluded that tonic fusimotor activity was present during unobstructed spontaneous stepping movements. In some cases this was predominantly dynamic and in other cases predominantly static. However, during moderately resisted movements tonic dynamic activity prevailed.

The filter technique to detect changes in dynamic sensitivity clearly has its limitations, since it relies on linear modelling of spindle properties, and since it was calibrated by recourse to spindle afferent responses to muscle stretching during tonic but not during modulated fusimotor activation. In view of the prominent non-linear characteristics of the muscle spindle (Sect. 3.1.3), in its present form this analytical technique can, therefore, only give approximative results.

The conclusions of *Prochazka* and *Wand* (1981) have since been confirmed by *Appenteng* et al. (1983) using a simulation technique to deduce (in acute preparations) the type and time course of fusimotor activity that shaped chronically recorded spindle afferent responses during voluntary movements (*Hulliger* and *Prochazka* 1983). An example is shown in Fig. 16, where the original response during an evoked step (Fig. 16A2) was satisfactorily reproduced (in an anaesthetized animal) by tonic stimulation of a dynamic γ-axon (Fig. 16C3). On the other hand, tonic stimulation of a static γ-fibre clearly failed, in this case, to imitate the prominent features of the original response (cf. Fig. 16C2 with 16A2). Yet for other examples of evoked or spontaneous stepping, evidence for tonic, or combined tonic and EMG modulated, static drive was obtained. Using such simulations, *Appenteng* et al. (1983) further corroborated the view that fusimotor activity could switch, for one and the same afferent, from predominantly static (during stepping) to predominantly dynamic (during imposed and moderately resisted stretch) (see also *Hulliger* et al. 1984).

The tentative conclusions which can be drawn from such analysis are: first, that during natural movements static and dynamic fusimotor neu-

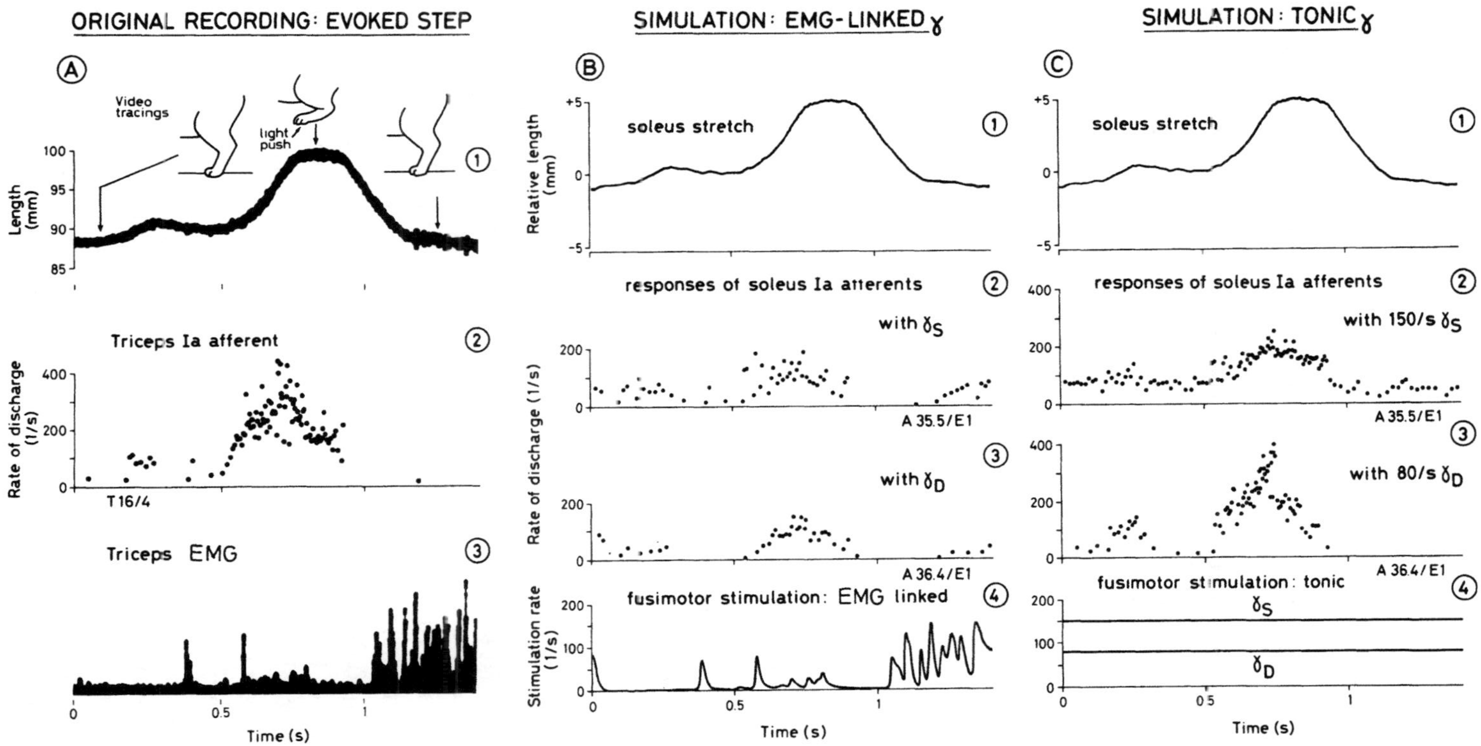

Fig. 16A–C. Original chronic recording (**A**) and simulations (**B**, **C**) of an evoked step. **A1** Video tracings of hindlimb and the monitored length of triceps. **A2** Instantaneous discharge rate of triceps (left gastrocnemius) Ia afferent. **A3** Rectified needle EMG from left gastrocnemius muscle. **B**, **C** Simulations with two separate Ia afferents (**B2**, **C2** and **B3**, **C3**) from soleus muscles of anaesthetized cats. **B4** EMG envelope used for rate-modulated γ-stimulation (**B2** and **B3**). **C4** Constant rates of γ-stimulation used in **C2** and **C3**. Responses in **B** and **C** shown as single-sweep-cycle histograms. Note that the best match of the original response (**A2**) is obtained with tonic γ_D stimulation. (From *Hulliger* and *Prochazka* 1983, with permission)

rones can be activated independently of each other (the choice depending on the type and/or context of movement) and second, that skeletomotor and fusimotor neurones can also be activated independently, since rhythmically alternating activity of the former was accompanied by tonic activity of the latter. Thus the activity in these three types of motoneurones need not be tightly coupled, suggesting that skeletofusimotor coactivation is not a general law.

In keeping with this view is the evidence for selective, and probably tonic, activation of dynamic fusimotor neurones supplying jaw-closing muscles during isometric biting in the monkey (*Larsen* et al. 1981). Under these conditions high-sensitivity (putative primary) units showed signs of moderate fusimotor bias and a very pronounced small-movement sensitivity, whereas presumed secondary afferents seemed largely unaffected by fusimotor drive. This clearly implicates dynamic but not static fusimotor neurones. However, the relative contribution of γ- and β-fibres cannot be assessed, given the likely activation during isometric contraction of slowly conducting dynamic β-motoneurones (see Sect. 3.5).

5.2.5 Absence of Strict α-γ Linkage?

In current usage, α-γ linkage is a special case of α-γ coactivation describing a high degree of similarity in the time course of skeletomotor and γ-fusimotor activity (see Sect. 6). In the extreme it entails direct proportionality (strict α-γ linkage). *Appenteng* et al. (1983), using the simulation technique described above (Sect. 5.2.4), examined whether the responses of hindlimb extensor afferents were attributable to such strict α-γ linkage. For the types of movement investigated the results speak a clear verdict. During both steps and imposed and resisted stretches, the chronically recorded responses could not be reproduced on stimulating either static or dynamic γ-motoneurones (Figs. 16B2, B3) when the time course of stimulation was rate-modulated by the EMG envelope (see Fig. 16B4). This was not astonishing for cases where the afferent response was sizable and stretch dominated and where at the same time there was EMG silence during stretching (as is the case for normal stepping; cf. Fig. 16). The finding is nevertheless in sharp contrast with the results obtained with comparable movements in reduced preparations (Sect. 5.1). On the other hand, for resisted stretches the failure of simulation of strict α-γ linkage was not necessarily predictable, since in this case afferent response and EMG envelope often were modulated in close sympathy.

Thus, in freely moving cats strict α-γ linkage does not occur for certain categories of movement. This, however, does not preclude its presence in other types of movement (cf. *Taylor* and *Appenteng* 1981) nor does it exclude various combinations of tonic background drive and superimposed EMG-modulated fusimotor activity. The general theme emerging from such findings is that fusimotor activity need not follow a single pattern. Instead, it may be flexibly adjusted to the particular requirements of each motor paradigm (see *Hulliger* et al. 1984).

5.3 Recordings from Spindle Afferents in Man

5.3.1 Human Spindles Studied In Vitro

Given the morphological similarity of muscle spindles in man and in other species (see Sect. 2.3), one should expect some degree of similarity in their functional properties. Studies on isolated human spindles have so far indeed supported this notion. Thus the frequency response characteristics of presumed primary and secondary afferents from finger extensor muscles (*Poppele* and *Kennedy* 1974) were indistinguishable from those seen in an identical in vitro preparation of cat tenuissimus spindles (*Poppele* and *Bowman* 1970). Likewise, in studies on spindle afferents from human intercostal muscles (*Newsom Davis* 1975), much the same values of absolute discharge rate, position sensitivity, and dynamic index were found as in the cat (*Harvey* and *Matthews* 1961; *Matthews* 1963; *Andersson* et al. 1968a; *Brown* et al. 1969b), provided that allowance was made for differences in muscle resting length (for details see *Prochazka* and *Hulliger* 1983). Moreover, like in the cat, spindle afferents with intermediate dynamic properties (see Sect. 3.1.1) were also encountered in man.

Further studies on human spindle properties in vitro seem highly desirable. In particular, it should be technically feasible to stimulate efferent axons or intrafusal muscle fibres to settle the question of whether a dual (static and dynamic) fusimotor supply of muscle spindles also exists in man (see Sect. 5.3.2.3).

5.3.2 Microneurography During Voluntary Motor Performance in Man

With the advent of the microneurographic recording technique (*Vallbo* and *Hagbarth* 1968; *Hagbarth* and *Vallbo* 1968) it became possible for the first time to record single-unit activity from peripheral nerves in man. Given the cooperation and flexible learning ability of human volunteers, a potentially much wider range of motor tasks has so been opened up for neurophysiological investigation, compared with the possibilities, for instance, or operant conditioning of laboratory animals.

The details of the technique have been described elsewhere (*Vallbo* 1972; *Vallbo* and *Hagbarth* 1973; *Vallbo* et al. 1979). At present its main shortcomings are the limitations in the classification of single units and in the stability of the recordings. The latter has restricted the range of speeds and amplitudes of movements investigated. Also the size of the voluntary contractions studied has been confined to the lower end of the physiological range of each muscle, partly because with larger voluntary efforts EMG signals from co-contracted muscles close to the recording site tend to interfere with the selectiveness of the single-unit recordings, and partly because of the risk of simply dislocating the recording electrode during vigorous contraction.

The problems encountered with the *classification* of afferent units are that maximum twitch tests to differentiate between spindle and Golgi tendon organ afferents often jeopardize further recording, and that primary and secondary spindle afferents cannot always reliably be distinguished: measurements of axonal conduction velocity remain to be routinely achieved, and succinylcholine is not applicable owing to the risks involved. Finally, classification of spindle afferents on the basis of their dynamic sensitivity to muscle stretch may be equivocal. (For a critical assessment of these matters see *Prochazka* and *Hulliger* 1983).

Multiunit recordings seem attractive as they might reveal ensemble response characteristics of selected groups of afferents. However, although they have been claimed to be dominated by activity in large Ia fibres (*Hagbarth* and *Young* 1979; *Burke* et al. 1979a), they are to be interpreted with extreme caution, given the likely risk of contamination by activity in the only slightly smaller Ib and α-axons (see *Prochazka* and *Hulliger* 1983).

5.3.2.1 Are Human Spindles Exceptional? Microneurographic recordings from spindle afferents in man stand out by the remarkably low discharge rates which have so far been encountered (for reviews see *Vallbo* et al. 1979; *Hagbarth* 1979; *D. Burke* 1981; *Vallbo* 1981). In relaxed muscle the afferents often are silent at short and intermediate muscle lengths, and at longer lengths peak discharge rates range between 10 and 25 impulses/s; also, peak rates during rapid stretch are often well below 100 impulses/s and, finally, similarly low discharge rates have mostly been found during voluntary contraction of the receptor-bearing muscle when fusimotor drive was present. All this is in striking contrast with the firing behaviour which is regularly encountered in acute or chronic recordings in cat and monkey (cf. Sect. 5.2.2).

The morphological evidence (Sect. 2.3) clearly suggestes that such scaled-down behaviour is hardly attributable to fundamental differences in structure *per se* between human and other mammalian spindles. However, it is conceivable that intramuscular topography provides an important scaling factor because, especially in long muscles, spindles need not lie in parallel (*Fulton* and *Pi Suñer* 1928) with the entire length of extrafusal muscle fibres. Instead, they often originate from, and insert at, such extrafusal fibres rather than the muscle tendon proper (see *Barker* 1974a), so that they are arranged in series with compliant elements. Given that human spindles are not substantially longer than other mammalian spindles (for references see Sect. 2.3), locally effective length changes during stretch might therefore constitute a substantially smaller fraction of the change in whole-muscle length than for the much shorter muscles of standard laboratory preparations.

Notwithstanding morphological considerations, it also seems very likely that the scaled-down behaviour of human spindle afferents is at least partly due to the much narrower range of experimental conditions which have so far been explored in microneurographic studies (see *Taylor* 1981; *Prochazka* and *Hulliger* 1983): the speeds of both active and passive movements as well as the size of voluntary contractions investigated have clearly been below the ranges which have routinely been covered in recordings from freely behaving animals. It appears therefore that in man neither the afferents' responsiveness nor the fusimotor efferents' excitatory strength has so far been tested to its full potential.

In addition to such quantitative aspects of spindle afferent behaviour, qualitative differences (between man and other mammals) concerning fusimotor strategy have also been emphasized. Above, it was a recurrent theme that in reduced preparations γ-motoneurones often revealed lower thresholds than α-motoneurones, both for central (Sect. 4.2.1) and reflex (Sect. 4.4.1) activation. Moreover, γ-motoneurones have repeatedly been seen to be spontaneously active when α-motoneurones were silent (*Hunt* 1951; *Hunt* and *Paintal* 1958; *Diete-Spiff* et al. 1962; *Ellaway* 1971, 1972; *Noth* 1971, 1983). Also, in alert cats there is indirect evidence for spontaneous fusimotor activity during EMG silence (*A. Prochazka,* personal communication). In line with these observations, *Burg* et al. (1974), *Szumski* et al. (1974) and *Struppler* and *Velho* (1976) reported on findings obtained in reflex studies and during Jendrassik's manoeuvre, which suggested that the same also held true in man. However, this view was challenged at once by *Hagbarth* et al. (1975a), and their critique has further been substantiated, especially on the basis of highly sensitive recordings designed to detect also minute amounts of EMG activity (*Burke* et al. 1976b, 1979a, 1980a,b). For the time being the bulk of the evidence strongly suggests that in the paradigms explored so far fusimotor neurones were only active when skeletomotor neurones also were (for detailed reviews see *Vallbo* et al 1979; *Hagbarth* 1979; *D. Burke* 1980, 1981). However, given the nature of the issue, the debate is likely to continue until a much wider range of motor paradigms has been explored with microneurographic recordings. For if descending control of γ-motoneurones in man were broadly similar to that in cat and monkey, then selective activation, especially of dynamic γ-motoneurones, must remain a distinct possibility of specific motor strategies.

5.3.2.2 Primary and Secondary Afferents

Dynamic Sensitivity and Variability of Discharge. In man, spindle afferent responses to stretch of the parent muscle may, on qualitative assessment, reveal either pronounced or modest dynamic sensitivity. Given the existence of primary and secondary *endings* (Sect. 2.3.1), such responses may reasonably be ascribed to primary and secondary *afferents.* Units with high dynamic sensitivity (primaries) tend to exhibit a more irregular resting discharge than units with low dynamic sensitivity (secondaries; *Burke* et al. 1979b; *Nordh* et al. 1983), which agrees with the properties of feline spindle afferents (*Matthews* and *Stein* 1969b). Similarly to the cat (*Brown* et al. 1967a), primaries in man are more prone to be driven by vibration than secondaries are (*Burke* et al. 1976a,b; *Roll* and *Vedel* 1982). However, whenever the parameters at issue were measured quantitatively, afferents with intermediate properties were also found (cf. Sect. 3.1.1), precluding their use as unequivocal criteria for unit classification (see references above; for dynamic index, *Hulliger, Thelin* and *Vallbo,* unpublished; also *Prochazka* and *Hulliger* 1983).

Small-Movement Sensitivity. Most investigators have observed that dynamically sensitive primary afferents tended to respond to small irregularities of load or movement with small bursts of impulses (*Vallbo* 1973a, 1974b, 1981; *Burke* et al. 1978a,b; *Hulliger* and *Vallbo* 1979b; *Hagbarth* and *Young* 1979; *Young* and *Hagbarth* 1980; *Hagbarth* et al. 1981; *McKeon* and *Burke* 1981, 1983). Such responses obviously were reminiscent of the striking small-movement sensitivity and the gain-compression property of cat Ia afferents (Sect. 3.1.2), yet the equivalence of these observations remains to be established in rigorous quantitative studies. In view of the scaled-down response properties of human spindle afferents (Sect. 5.3.2.1), their small-movement sensitivity may turn out to be less impressive, quantitatively, and less important, functionally, than in the cat.

5.3.2.3 Fusimotor Control

γ-or β-Action? Evidence has been obtained that during voluntary contraction human spindles are subjected to fusimotor drive. This rests on three types of observations. First, in the absence of movement spindle discharge has consistently been found to increase with increasing isometric contraction force (*Vallbo* 1970a,b, 1971, 1972, 1973b, 1974b; *Hagbarth* et al. 1975a; *Burg* et al. 1976; *Burke* et al 1976b, 1978c, 1979a, 1980a,b). Second, during slow active lengthening movements spindle discharge was higher than when the same or similar movements were performed passively (*Burke* et al. 1978b; *Hulliger* and *Vallbo* 1979a; *Roll* and *Vedel* 1982). Third, during slow shortening movements spindle afferent firing was found to persist or even increase, whilst during passive shortening especially primary afferents tended to fall silent (*Vallbo* 1973b; *Hagbarth* et al. 1975b; *Burke* et al. 1978b; *Hulliger* and *Vallbo* 1979a, and unpublished; *Roll* and *Vedel* 1982). Whilst these findings are most convincingly interpreted as manifestations of fusimotor activity, on their own they do not permit definite conclusions concerning the relative amounts of β- or γ-activity involved.

Burke et al. (1979a), using *Leksell's* (1945) pressure block technique, have shown that at least part of the fusimotor effects observed during voluntary contraction in man are mediated by small γ-motoneurones, since for half of their observations fusimotor excitation of spindle afferents persisted during a complete block of α- and therefore also of β-motoneurones. The additional finding that in the remaining cases spindle excitation was no longer present during a voluntary effort (once the α-block was complete) is most reasonably interpreted as evidence for an involvement also of β-fusimotor fibres (for details see *Prochazka* and *Hulliger* 1983). This interpretation clearly is in agreement with the wealth of morphological evidence for the existence of β-innervation in general (Sect. 2.1.5) and in man (Sect. 2.3.1). Whilst the findings of *Burke* et al. (1979a) were obtained under restricted (isometric) conditions, there is at present no compelling reason to assume that the balance between γ- and β-contribution to overall fusimotor action should be fundamentally different in the more general case of voluntary load-bearing movements. In conclusion, fusimotor effects, as soon as they are accompanied by skeletomotor activity, are likely to be mediated by both β- and γ-motoneurones, yet there is at present no simple technique to asses their relative contributions.

Static and Dynamic Action? Given the existence of bag_1 fibres in human spindles (Sect. 2.3.1) and of static and dynamic neurones in monkey (*Koeze* 1968; *Cheney* and *Preston* 1976b,c), one should also expect to find separate static and dynamic fusimotor effects in man. However, the crucial tests, relying on the observation of decreased or increased dynamic sensitivity in responses to sufficiently large stretches (see Sect. 3.2.1), have not so far been reported (cf. *Vallbo* 1981; *Hulliger* 1981). The preliminary findings of *Burg* et al. (1976) are difficult to assess in the absence of quantitative analysis.

If secondary afferents in man were also selectively operated by static fusimotor neurones (see Sect. 2.2.1), then static fusimotor contribution to spindle excitation could be inferred from a number of studies where primary and secondary afferents were differentiated (*Vallbo* 1971, 1972, 1974b; *Hagbarth* et al. 1975a; *Burke* et al. 1976b; *Hulliger* and *Vallbo* 1979a; *Vallbo* and *Hulliger* 1982; *Hulliger* et al. 1982). Clearly, studies on isolated spindles could here greatly clarify matters (cf. Sect. 5.3.1).

In contrast, Ia excitation during active muscle shortening cannot automatically be attributed to static fusimotor action, since for sufficiently slow movements dynamic γ-axons are quite capable of maintaining Ia firing (see Fig. 8C), especially when shortening is accompanied by oscillatory disturbances (cf. Fig. 12C, and *Baumann* et al. 1982).

5.3.2.4 Skeletofusimotor Coactivation. Soon after the recognition of the fusimotor role of γ-motoneurones (see Sect. 2.1.2) the theory was proposed that certain movements were controlled by the stretch reflex operated as a follow-up servo system (*Merton* 1951, 1953; see also *Rossi* 1927). According to this concept, especially slow movements were thought to be initiated by indirect activation of α-motoneurones via γ-loop and monosynaptic excitation by Ia afferents rather than by direct descending commands. If this were true the activation of γ-motoneurones and spindle afferent fibres should precede the onset of EMG activity at the beginning of a voluntary contraction. It was one of the most influential results of microneurographic research so far that it could unequivocally be shown that this was not the case. *Vallbo* (1971) found that at the onset of a voluntary isometric contraction the excitation of spindle afferents invariably lagged behind the first burst of EMG activity, indicating that at best fusimotor neurones were activated at the same time as, but not before, α-motoneurones were. This finding has since been repeatedly

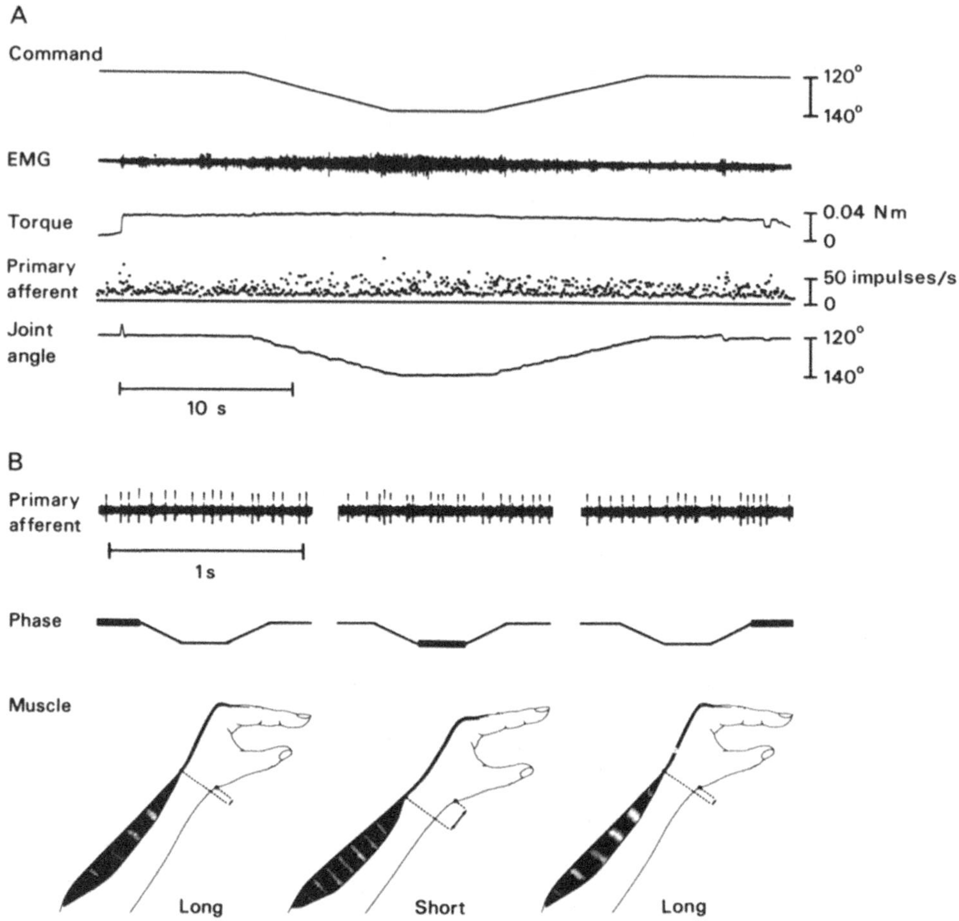

Fig. 17A,B. Tracking movement performed by human subject with the fourth finger of the left hand. **A** From *top* to *bottom,* time course of command signal (extension movements downwards); EMG record as obtained with surface electrodes; torque; activity of a primary afferent unit from the extensor digitorum communis muscle (fourth finger portion) displayed as instantaneous rate of discharge; and metacarpophalangeal joint angle of the fourth finger. **B** Sample records of the same afferent unit's discharge during 1 s periods of the three separate hold phases and schematic illustration of the metacarpophalangeal joint angle and muscle length of the finger extensors during these hold phases. Finger extension (*short*) was accompanied by muscle shortening and is plotted downwards. (From *Hulliger* et al. 1982, with permission)

confirmed (*Hagbarth* et al. 1975b; *Burke* and *Eklund* 1977; *Burke* et al. 1978b,c, 1980a,b).

The dominating theme since then has been that voluntary contraction in man is accompanied by "fusimotor outflow which largely runs parallel with the skeletomotor outflow in space, time and intensity" (*Vallbo* 1973b), by "tight α-γ coactivation" (*Burke* et al. 1979a), or by "α-γ

coactivity and α-γ cosilence" (*Hagbarth* 1981) (see also *Vallbo* et al. 1979; *Hagbarth* 1979; *D. Burke* 1980, 1981; *Vallbo* 1981; and above, Sect. 5.3.2.3). The example of Fig. 17 shows the response of a primary afferent during a load-bearing tracking movement performed with the fourth finger. The afferent was not silenced during the shortening movement (plotted downwards), whilst it paused when similar movements were imposed passively (not illustrated). Also, during the subsequent lengthening phase there was no conspicuous dynamic response, as was present during passive movements. Thus, modulated fusimotor activity just adequate to offset the afferent's length response probably accompanied the voluntary contraction of Fig. 17, and the time course of fusimotor outflow may indeed have been rather similar to that of skeletomotor activity, as seen in the EMG record.

For isometric contractions the notion of skeletofusimotor coactivation describes reasonably uniform spindle behaviour, i.e. an acceleration of spindle afferent firing at fixed muscle length which tends to increase with the strength of contraction (see Sect. 5.3.2.3). In contrast, for alternating movements the afferent discharge patterns which have been ascribed to coactivation or even tight linkage are highly non-uniform: depending on the speed of movement anything from length-independent discharge or even acceleration during shortening (slow movements, *Hulliger* and *Vallbo* 1979a; *Hulliger* et al. 1982; Fig. 17) to length-dominated behaviour (faster movements, *Burke* and *Eklund* 1977; *Burke* et al. 1978b) has been encountered. Between these extremes there are bimodal responses showing transient spindle excitation both during stretch and during shortening (*Hagbarth* et al. 1975b; *Burke* et al. 1978b; *Roll* and *Vedel* 1982).

These observations merit attention for two reasons: First, it may be argued that in spite of such non-uniformity in afferent behaviour the underlying pattern of fusimotor activity still was uniform and closely linked to skeletomotor activity, since such variety should indeed be expected from the interaction of a fixed pattern of fusimotor discharge with movements of widely ranging speeds and amplitudes (see Sect. 5.2.3). However, if this were true, the concept of linkage would loose its putative purpose, since linked fusimotor discharge would then no longer be capable of providing an ongoing working discharge of spindle afferents for a wide range of movements (see *Granit* 1970, and Sect. 1). Second, uniformly linked fusimotor behaviour would implicate β-, as well as static and dynamic γ-, motoneurones (provided that they existed in man, cf. Sect. 5.3.2.3) with all of them exhibiting EMG-like discharge patterns. Yet qualitative interpretation of afferent recordings hardly permits such deductions, since by the same token tonic or rectangular profiles of fusimotor activity, perhaps combined with some α-linked admixture, could then be invoked to account for some of the responses described above.

Clearly, such possibilities must be borne in mind in view of the quantitative estimates of fusimotor outflow which have been obtained for chronic cat data (*Appenteng* et al. 1983; see Sects. 5.2.4, 5.2.5), and which demonstrated the presence of tonic fusimotor drive in afferent responses not unlike some of those seen during alternating movements in man.

In conclusion, it must be seriously considered that what has been subsumed under 'coactivation' in fact were weighted mixtures of tightly linked β-activity and widely ranging patterns of either static or dynamic γ-activity. Simulation studies, as outlined above (Sect. 5.2.4), may help specify such patterns, yet 'coactivation' as a general describing term is more likely to blur than to define the fusimotor strategies accompanying such alternating movements.

Orderly Recruitment of Fusimotor Neurones? In isometric contractions spindle afferents were invariably found to be activated at fixed levels of contraction strength (*Burke* et al. 1978c). From this observation it was concluded that the activation of γ-motoneurones was governed by a fixed order, as is the case for α-motoneurones (see Sect. 3.5). However, as was detailed elsewhere (*Prochazka* and *Hulliger* 1983), this interpretation must be regarded as highly speculative, simply because the evidence to support it is limited and rather indirect. Recruitment of β-motoneurones is likely to have substantially contributed to these findings and, at best, the evidence for such orderly recruitment applies to static γ-motoneurones, given the modest excitatory strength of dynamic γ-fibres at fixed muscle length. Such reservations are further borne out by the direct demonstration of *Murphy* (1981) that in the rabbit a fixed recruitment order amongst pairs of γ-motoneurones was not a general feature.

5.3.2.5 Position Coding. Following a protracted debate (for references see *McCloskey* 1978, 1981; *Matthews* 1982) concerning the contribution of muscle spindle afferents to position sense and kinaesthesia, *Goodwin* et al. (1972) made out a convincing case for such a sensory role of these proprioceptors. The evidence to support this view was derived from striking illusions of movement which were induced by muscle vibration and which strongly implicated primary spindle afferents. However, given the complex interactions of changing muscle length and (perhaps modulated) fusimotor drive during voluntary movement, it was by no means self-evident that the sensory message conveyed by spindle afferents was an easy-to-read code containing overt components of position, velocity, or higher-order derivative signals.

The responses of *passive* spindle afferents contain both static (position) and dynamic (velocity, acceleration) components, and passive position sensitivity of both primary and secondary afferents in man is of the same order of magnitude as in cat hindlimb muscles, provided that allowance is made for muscle resting length (*Vallbo* 1974a; *Hulliger* et al. 1982). The same probably holds true for dynamic responses, although this remains to be quantified. However, during *active* position holding, position responses were no longer detectable (*Hulliger* et al. 1982), suggesting that under these conditions fusimotor drive was scaled so as to offset the effects of changing muscle length (see also Fig. 17). Preliminary evidence further suggests that during very slow voluntary movements dynamic responses are minimal and probably insignificant (*Hulliger, Nordh* and *Vallbo,* to be published). During faster alternating movements, responses to dynamic muscle stretch are regularly seen, yet again, the message is far from straightforward (see bimodal responses, Sect. 5.3.2.4).

How then can faithful kinaesthetic information be extracted from such complex afferent signals from the prime movers? There are at least two likely mechanisms which might well act cooperatively. First, the CNS might have recourse to corollary discharge of fusimotor outflow to compute the time course of movement, performing in essence the inverse operation of the peripheral convolution of muscle length and fusimotor drive. Albeit this seems feasible, the task would nevertheless be formidable, given the intricate non-linearities of spindle behaviour. Second, antagonist muscles, provided that they are not co-contracted, are bound to provide much less complicated, but still non-linear, monitoring of the movement accomplished by the agonists (prime movers) (*Roll* and *Vedel* 1982). On the evidence of *Goodwin* et al. (1972), there is indeed no reason to deny these signals access to consciousness. (For more detailed discussion of this issue see *McCloskey* 1978, 1981; *Matthews* 1982.)

5.3.2.6 α-γ-Independence. So far, little evidence has been adduced in studies in man to show that fusimotor activity may be controlled independently of skeletomotor activity, since until recently the prevailing view was that fusimotor and skeletomotor activity were linked (see Sect. 5.3.2.4). However, preliminary results indicate that under circumscribed conditions of motor performance the balance between skeletomotor and fusimotor activity may not be rigidly fixed (*Vallbo* and *Hulliger* 1979, 1981; *Burke* et al. 1980b).

Burke et al. (1980b) found that in isometric contractions the threshold of activation of fusimotor neurones (as deduced from spindle afferent discharge) could be altered by various manoeuvres designed to alter presumed descending or segmental reflex inputs to these neurones. *Vallbo* and *Hulliger* (1979, 1981) observed short-lasting periods of excitation of primary spindle afferents at the beginning and at the end of auxotonic (but not of isotonic) tracking movements, which were not accompanied by corresponding transients in EMG activity and which were not explicable in terms of extrafusal mechanical events.

The most straightforward interpretation of both sets of findings is that fusimotor efferents (most likely γ-motoneurones) were activated from 'private' input pathways. Obviously, these studies provide but indirect evidence for α-γ independence, and the data available do not permit differentiation between static and dynamic fusimotor contributions. Nevertheless, they may indicate the lines which ought to be explored in order to further delimit the validity of the notions of α-γ linkage and α-γ independence.

In this context the case of exclusive α-activation may also be named. During both the tonic vibration reflex (*Burke* et al. 1976b) and the tendon jerk (*Burg* et al. 1973; *Szumski* et al. 1974; *Hagbarth* et al. 1975c; *Wallin* and *Hagbarth* 1978) spindle afferent behaviour was indicative of fusimotor silence, providing another example of α-γ independence. These findings are hardly surprising, since both reflexes are mediated largely by Ia fibres, which act preferentially on α-motoneurones (see Sect. 4.4.2).

5.3.3 Future Directions

In man the present picture shows spindle afferents discharging at rather low and unimpressive rates during voluntary contractions, and it may be inferred that fusimotor neurones are not strongly activated either. As regards the pattern of fusimotor discharge, the notion has dominated the field that for most of the motor paradigms investigated fusimotor and skeletomotor neurones were coactivated or even linked.

These findings and interpretations need not be the final truth. The concept of coactivation may relax its grip on microneurographers when a wider range of motor tasks will be investigated, including, e.g. imposed movements which are resisted by the subject, novel motor tasks whilst they are gradually acquired by the subject, or movements which have to be performed without the customary aid of visual feedback. Also, it may turn out that coactivation will no longer be an adequate describing term when evidence for separate static or dynamic fusimotor activity during various movements will be searched for (see Sect. 5.3.2.3). Whilst it is probably unrealistic to expect, in the foreseeable future, direct recordings from identified and classified γ-motoneurones, indirect but quantitative methods may nevertheless lead to important new insights.

6 Conclusions

The debate surrounding the role of fusimotor neurones during voluntary movement has been fuelled by three types of problems. First, in the absence of direct recordings from γ-motoneurones, qualitative deductions from afferent recordings have been more influential than is desirable; second, loose and insufficiently standardized terminology has helped to confuse the issue; and third, the unspoken expectation of a unifying theory may have unnecessarily favoured dogmatism.

Considering the last aspect first, there seems at present to be no compelling reason to uphold a *single strategy of fusimotor control* for all types of

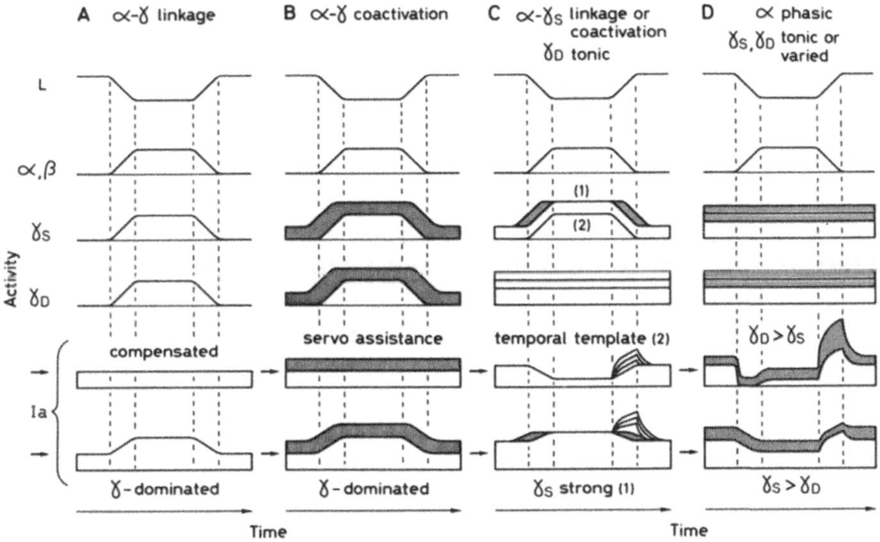

Fig. 18A–D. Schematic display of α-, β-, γ-efferent and spindle primary afferent firing rates during idealized voluntary movements, for different strategies of fusimotor control. *Top* to *bottom:* muscle length (*L*) shortening downwards; α-, β-, γ_S-, γ_D-firing rates; Ia firing rates (*upper* assuming weak γ-action, *lower* assuming strong γ-action). **A** α-γ Linkage (extreme case: strict proportionality between α- and γ-firing rates). **B** α-γ Coactivation (*shaded areas* indicate the range of possibilities). **C** α-γ_S Coactivation, γ_D-tonic (*1:* γ_S-coactivation strong; *2:* γ_S-modulation a 'temporal template' of the length changes). Note increasing levels of tonic γ_D-firing rates, reflected in increasing Ia responses during muscle lengthening. **D** Independence of both γ_S- and γ_D-activation from α- and β-activation (special case: constant γ_S- and γ_D-firing rates, phasic α- and β-activation). (From *Prochazka* and *Hulliger* 1983, with permission)

voluntary movement. The complexity of both descending and reflex control of γ-motoneurones is such that there must be considerable scope for choice. Thus, the concepts discussed below should be viewed as equivalent options from a wide *spectrum of possibilities* rather than as exclusive alternatives. The aim of future research then must be to define more clearly the particular type, time course and intensity of fusimotor outflow for a given category of movement.

Loose usage of a given term may reflect its inherent limitations of precision. Apart from the simplest cases, it seems unlikely that words will ever be apt to define the complex features of motor activity. Instead, the use of simple graphic displays showing representative profiles (cf. Fig. 18) or some quantitative assessment of the degree of similarity of different patterns of activity may eventually help overcome this problem. In spite of such reservations, *Prochazka* and *Hulliger* (1983) have attempted to delimit the range of possibilities encompassed by the notions of 'linkage', 'coactivation', and 'independence'. These definitions are briefly recapitulated here to avoid ambiguity in the following discussion.

The concept of 'α-γ *linkage*' (Fig. 18A) was originally based on the finding that "α- and γ-reflexes have proved to be linked, co-excited and co-inhibited, often with the γ-reflexes leading" (*Granit* 1955, p. 254). 'Linkage' may therefore be used to describe a close similarity in the time course of α- and γ-activity when, apart from an earlier recruitment of γ-motoneurones, these are only active when α-motoneurones also are. The virtue of this formulation lies in its relative precision. Its limitation is that short of direct recordings from α- and γ-motoneurones its presence is difficult to ascertain.

In current usage 'α-γ *coactivation*' is a looser term implying some, but not necessarily a tight, coupling between α- and γ-activity. The range of possibilities is indicated by the shaded areas of Fig. 18B which delimit patterns of static and dynamic γ-activity. In particular 'coactivation', on this definition, encompasses patterns of γ-activity where a phasically modulated and broadly EMG-linked component is superimposed on tonic background discharge. Coactivation also lies behind the concept of *servo* assistance, where fusimotor outflow is adjusted so as to compensate for spindle unloading during shortening contractions and to maintain spindle firing at an approximately constant level (*Matthews* 1972; *Stein* 1974; and Fig. 17).

The term 'α-γ *independence*' is used to describe a wide range of patterns of γ-activity, whose common feature is an even higher degree of deviation from the profile of α-activity (see examples of Fig. 18C,D). Thus, rhythmically modulated α-activity might be accompanied by tonic (i.e. non-modulated) γ-activity, either restricted to dynamic (Fig. 18C) or extending to both dynamic and static (Fig. 18D) γ-motoneurones. Numerous other patterns, including profiles of γ-activity which are modulated but out of phase with α-activity, would also fall into this category.

The bottom diagrams of Fig. 18 show patterns of primary afferent discharge which may result from the interaction of a given constellation of β- and γ-activity with changing muscle length. For components of γ-activity modulated in phase with α-activity, such interaction may lead to various degrees of compensation of spindle unloading ranging from hardly noticeable (length-dominated response) to overcompensated (γ-dominated response), depending on the degree of mismatch between intrafusal and extrafusal length change (bottom diagrams, cf. upper with lower traces, i.e. weak with strong γ-action). The responses illustrated probably apply only to relatively slow movements. For fast movements, EMG-modulated components of fusimotor activity stand little chance of exerting themselves (see Sect. 5.2.3), whereas tonic fusimotor drive, especially if it is dynamic, should bring stretch dominated afferent behaviour into greater prominence than is illustrated in Fig. 18.

The range of fusimotor strategies illustrated in Fig. 18 probably encompasses most of the experimental evidence available and most of the concepts which are debated at present. The following may be taken as a brief summary of current opinion.

α-γ *Linkage* has been demonstrated in reduced preparations (Sect. 5.1) for locomotion and in respiratory muscles for those γ-motoneurones which were rhythmically modulated in phase with α-activity (Sect. 5.1). The least ambiguous example of (*skeletofusimotor*) *linkage* is of course the invariant component of fusimotor modulation provided by β-motoneurones.

α-γ *Coactivation* may be the adequate describing term for the bulk of the recordings from human subjects. But this assignment reflects uncertainty of interpretation more than uniformity of fusimotor behaviour. As detailed above (Sect. 5.3.2.4), more quantitative analysis may single out cases of tighter linkage or of greater independence. The same applies for chronic afferent recordings during jaw movements in monkey (*Goodwin* and *Luschei* 1975), as well as for some of the recordings from freely moving cats (*Loeb* and *Duysens* 1979; *Loeb* and *Hoffer* 1981).

α-γ *Independence* of varying degree has mainly been inferred from recordings in cats, but some evidence for its occurrence has also been obtained in man (Sect. 5.3.2.6). In reduced cat preparations this applies especially to the observation of tonic γ-firing during rhythmic α-activation (respiration, mastication; see Sect. 5.1). *Matthews* (1972) first suggested that α-γ linkage might be restricted to static γ-fibres, leaving dynamic γ-axons to be controlled more independently. Expanding on this theme for the case of jaw movements, *Taylor* and *Appenteng* (1981) formulated the concept of static γ-action providing a 'temporal template' of the intended movement (α-γ_S coactivation) and of tonic dynamic γ-action simultaneously controlling incremental sensitivity (see Fig. 18C). Finally *Prochazka* and colleagues (Sect. 5.2.4) emphasized, on the evidence of chronic recordings from freely moving cats, an even higher degree of α-γ independence, with largely tonic behaviour of both static and dynamic γ-motoneurones during rhythmically modulated α-activity (Fig. 18D).

As to the question of 'linkage or independence', the final answer may be 'neither, nor. . .', and the examples of Fig. 18 may be nothing more than special cases in a continuous spectrum of control strategies.

The concept of *servo assistance* featured in its specific formulation (*Matthews* 1964, 1972; *Stein* 1974; Fig. 18B) a particular form of coactivation. On this view fusimotor action (rather than activity) should reflect the time course of an intended movement so as to largely offset spindle afferent responses to shortening and lengthening. In the extreme case spindle discharge during undisturbed movement would be maintained at a steady level (*Phillips* 1969). For most movements this would have to be accomplished by static fusimotor action, leaving an extra degree of freedom for the control of dynamic γ-motoneurones. Thus, in its wider formulation and especially when the requirement of strict constancy of afferent discharge is relaxed (see *Matthews* 1981b, for details), servo assistance might also be provided by a number of independent strategies (Fig. 18C,D). In particular, it would then enclose the concept of the γ_S temporal template (Fig. 18C). Whilst examples of servo assistance type of afferent behaviour have often been encountered (see, e.g. Fig. 17), it is also evident that the concept hardly applies to fast movements, since these so far have revealed length-dominated afferent behaviour (Sect. 5.2.3). Finally, it may be noted that most of the concepts and afferent response patterns illustrated in Fig. 18 are variations of the theme of maintenance of spindle feedback during muscle shortening, which was introduced by *Hunt* and *Kuffler* (1951,a; *Kuffler* et al. 1951). The refinements which have since been added stem from the appreciation of the importance of the speed of movement and of the time course of fusimotor outflow, as well as from the recognition of the separate roles that may be played by static and dynamic fusimotor neurones.

The CNS having two independent channels (γ_S, γ_D) to control Ia afferent firing during movement, a particular pattern of response (Fig. 18, bottom graphs) may within limits be brought about by a number of different combinations of γ_S- and γ_D-output profiles. This then provides scope for *parameter control* of Ia sensitivity, which may be achieved by altering the balance between static and dynamic fusimotor outflow (*Hulliger* et al. 1977b; *Matthews* 1981b,c; see Sect. 3.2.4, combined action). Given the dual nature of the γ-fusimotor system and the evidence for independent central control (Sects. 4.2 and 4.4), this must be considered as one of the key functions of this system. Examples are shown in the patterns of afferent discharge of Fig. 18C and D, where the size of dynamic response (controlled parameter) varies with the relative strength of dynamic, compared with static, action. However, the characteristics of such parameter control are far from simple. This is best illustrated by its failure where it might be most attractive, namely, in the graded control of small-movement sensitivity. This is due to the occlusion of dynamic by static action in Ia responses to small movements (see Sect. 3.2.4).

Why should skeletomotor and fusimotor neurones be controlled independently? Until we know to what extent this is indeed the case and until we know its benefits for motor performance, we cannot tell with certainty. All we know is that α- and γ-motoneurones have evolved as separate entities in mammals and that the potential for differential control of their activity is laid down in the brain. *Why* then should they *not* be controlled independently? The debate is teleologic and, in isolation, serves little purpose. Yet, if we accept present evidence at face value, then patterns ranging from 'linkage' to 'independence' provide a choice of different fusimotor strategies whose functional significance, though, remains to be fully elucidated.

Acknowledgements. This work was supported by grants from the Swiss National Science Foundation (No. 3.225.77, 3.157.81, and 3.228.82), the Jubiläumsstiftung of the University of Zürich, the Dr. Eric Slack-Gyr Foundation and the European Science Foundation (twinning grant Nos. 553, 1145). I should like to thank Dr. P.B.C. Matthews, Dr. J. Noth and Dr. P. Zangger for valuable comments on the manuscript, and I am greatly indebted to Ms. R. Emch, Ms. E. Schneider and Mr. H. Künzli for excellent technical assistance, and to Ms. B. Hofer for patient typing. Thanks are extended to Drs. I.A. Boyd, G.M. Goodwin, E.S. Luschei, P.B.C. Matthews. A. Prochazka, R.F. Schmidt and P. Wand for permission to reproduce Figs. 2,3,4,14,15 and Tables 1 and 2.

References

Abdullah A, Eldred E (1959) Activity in gamma efferent circuits induced by distension of the bladder. Neuropath Exp Neurol 18:590–596

Abrahams VC (1981) Sensory and motor specialization in some muscles of the neck. TINS 4:24–27

Adal MN, Barker D (1962) Intramuscular diameters of afferent nerve fibres in the rectus femoris muscle of the cat. In: Barker D (ed) Symposium on muscle receptors. Hong Kong University Press, Hong Kong, pp 249–256

Adal MN, Barker D (1965) Intramuscular branching of fusimotor fibres. J Physiol 177:288–299

Adrian ED, Zotterman Y (1926) The impulses produced by sensory nerve endings. part 2. The response of a single end-organ. J Physiol 61:151–171

Agnew RF, Preston JB (1965) Motor cortex-pyramidal effects on single ankle flexor and extensor motoneurones of the cat. Exp Neurol 12:384–398

Agnew RF, Preston JB, Whitlock DG (1963) Patterns of motor cortex effects on ankle flexor and extensor motoneurones in the "pyramidal" cat preparation. Exp Neurol 8:248–263

Alnaes E, Jansen JKS, Rudjord T (1964) Fusimotor activity in the spinal cat. Acta Physiol Scand 63:197–212

Andén NE, Jukes MGM, Lundberg A, Vycklický L (1966a) The effect of DOPA on the spinal cord. 1. The influence on transmission of primary afferents. Acta Physiol Scand 67:373–386

Andén NE, Jukes MGM, Lundberg A (1966b) The effect of DOPA on the spinal cord. 2. A pharmacological analysis. Acta Physiol Scand 67:387–397

Andersson BF, Lennerstrand G, Thoden U (1968a) Fusimotor effects on position and velocity sensitivity of spindle endings in the external intercostal muscle of the cat. Acta Physiol Scand 74:285–300

Andersson BF, Lennerstrand G, Thoden U (1968b) Response characteristics of muscle spindle endings at constant length to variations in fusimotor activation. Acta Physiol Scand 74:301–318

Andrew BL, Part NJ (1974) The division of control of muscle spindles between fusimotor and mixed skeletofusimotor fibres in rat caudal muscle. Q J Exp Physiol 59: 331–349

Andrew BL, Leslie GC, Part NJ (1978) Some observations on the efferent innervation of rat soleus muscle spindles. Exp Brain Res 31:433–443

Andrew BL, Leslie GC, Part NJ (1979) Some observations on responses evoked by pinna stimulation in efferent and afferent fibres innervating hindlimb muscle spindles of the rat. Exp Brain Res 34:263–272

Unless otherwise stated "J Physiol" refers to J Physiol (Lond)

Appelberg B (1981) Selective central control of dynamic gamma motoneurones utilised for the functional classification of gamma cells. In: Taylor A, Prochazka A (eds) Muscle receptors and movement. MacMillan, London, pp 97–107

Appelberg B, Emonet-Dénand F (1965) Central control of static and dynamic sensitivities of muscle spindle primary endings. Acta Physiol Scand 63:487–494

Appelberg B, Kozary IZ (1963) Excitation of flexor fusimotor neurones by electrical stimulation in the red nucleus. Acta Physiol Scand 59:445–453

Appelberg B, Molander C (1967) A rubro-olivary pathway. I. Identification of a descending system for control of the dynamic sensitivity of muscle spindles. Exp Brain Res 3:372–381

Appelberg B, Bessou P, Laporte Y (1966) Action of static and dynamic fusimotor fibres on secondary endings of cat's spindles. J Physiol 185:160–171

Appelberg B, Jeneskog T, Johansson H (1975) Rubrospinal control of static and dynamic fusimotor neurones. Acta Physiol Scand 95:431–440

Appelberg B, Johansson H, Kalistratov G (1977) The influence of group II muscle afferents and low threshold skin afferents on dynamic fusimotor neurones to the triceps surae of the cat. Brain Res 132:153–158

Appelberg B, Hulliger M, Johansson H, Sojka P (1979) Excitation of dynamic fusimotor neurones of the cat triceps surae by contralateral joint afferents. Brain Res 160:529–532

Appelberg B, Hulliger M, Johansson H, Sojka P (1981) Reflex activation of dynamic fusimotor neurons by natural stimulation of muscle and joint receptor afferent units. In: Taylor A, Prochazka A (eds) Muscle receptors and movement. MacMillan, London, pp 149–161

Appelberg B, Hulliger M, Johansson H, Sojka P (1982a) Fusimotor reflexes in triceps surae elicited by natural stimulation of muscle afferents from the cat ipsilateral hindlimb. J Physiol 329:211–229

Appelberg B, Hulliger M, Johansson H, Sojka P (1982b) An intracellular study of rubrospinal and rubro-bulbospinal control of lumbar γ-motoneurones. Acta Physiol Scand 116:377–386

Appelberg B, Hulliger M, Johansson H, Sojka P (1983a) Actions on γ-motoneurones elicited by electrical stimulation of muscle group I afferent fibres in the hindlimb of the cat. J Physiol 335:237–253

Appelberg B, Hulliger M, Johansson H, Sojka P (1983b) Actions on γ-motoneurones elicited by electrical stimulation of muscle group II afferent fibres in the hindlimb of the cat. J Physiol 335:255–273

Appelberg B, Hulliger M, Johansson H, Sojka P (1983c) Actions on γ-motoneurones elicited by electrical stimulation of muscle group III afferent fibres in the hindlimb of the cat. J Physiol 335:275–292

Appelberg B, Hulliger M, Johansson H, Sojka P (1983d) Recurrent actions on γ-motoneurones mediated via large and small ventral root fibres in the cat. J Physiol 335: 293–305

Appenteng K, Morimoto T, Taylor A (1980) Fusimotor activity in masseter nerve of the cat during reflex jaw movements. J Physiol 305:415–431

Appenteng K, Lund JP, Seguin JJ (1982a) Behavior of cutaneous mechanoreceptors recorded in mandibular division of gasserian ganglion of the rabbit during movements of the lower jaw. J Neurophysiol 47:151–166

Appenteng K, Prochazka A, Proske U, Wand P (1982b) Effect of fusimotor stimulation on Ia discharge during shortening of cat M. soleus at different speeds. J Physiol 329:509–526

Appenteng K, Hulliger M, Prochazka A, Zangger P (1983) Fusimotor action during movement, deduced by matching the patterns of afferent discharge in freely moving and anaesthetized cats. J Physiol 334:29–30P

Arbuthnott ER, Ballard KJ, Boyd IA, Gladden MH, Sutherland FI (1982) The ultrastructure of cat fusimotor endings and their relationship to foci of sarcomere convergence in intrafusal fibres. J Physiol 331:285–309

Ariano MA, Armstrong RB, Edgerton VR (1973) Hindlimb muscle fibre populations of five mammals. J Histochem Cytochem 21:51–55
Bakker GJ, Richmond FJR (1981) Two types of muscle spindles in cat neck muscles: a histochemical study of intrafusal fibre composition. J Neurophysiol 45:973–986
Bakker DA, Richmond FJR (1982) Muscle spindle complexes in muscles around upper cervical vertebrae in the cat. J Neurophysiol 48:62–74
Baldissera F, Hultborn H, Illert M (1981) Integration in spinal neuronal systems. In: Brookhart JM, Mountcastle VB, Brooks VB (eds) Motor control, part 1. American Physiological Society, Bethesda, pp 509–595 (Handbook of physiology, Sect 1, vol 2)
Ballard KJ (1978) Typical sympathetic noradrenergic endings in a muscle spindle of the cat. J Physiol 285:61–62P
Banks RW (1971) Histochemical studies on rabbit intrafusal fibres. J Anat 108:613–614
Banks RW (1981) A histological study of the motor innervation of the cat's muscle spindle. J Anat 133:571–591
Banks RW, Barker D, Stacey MJ (1977a) Intrafusal branching and distribution of primary and secondary afferents. J Physiol 272:66–67P
Banks RW, Harker DW, Stacey MJ (1977b) A study of mammalian intrafusal muscle fibres using a combined histochemical and ultrastructural technique. J Anat 123:783–796
Banks RW, Barker D, Bessou P, Pagès B, Stacey MJ (1978) Histological analysis of cat muscle spindles following direct observation of the effects of stimulating dynamic and static motor axons. J Physiol 283:605–619
Banks RW, Barker D, Stacey MJ (1979) Sensory innervation of cat hindlimb muscle spindles. J Physiol 293:40–41P
Banks RW, Barker D, Stacey MJ (1981a) Structural aspects of fusimotor effects on spindle sensitivity. In: Taylor A, Prochazka A (eds) Muscle receptors and movement. MacMillan, London, pp 5–16
Banks RW, Ellaway PH, Scott JJ (1981b) Responses of de-efferented muscle spindles of peroneus brevis and tertius muscles in the cat. J Physiol 210:53P
Banks RW, Barker D, Stacey MJ (1982) Form and distribution of sensory terminals in cat hindlimb muscle spindles. Philos Trans R Soc Lond (Biol) 299:329–364
Barany M (1967) ATPase activity of myosin correlated with speed of muscle shortening. J Gen Physiol (Suppl) 50:197–216
Barker D (ed) (1962a) Symposium on muscle receptors. Hong Kong University Press, Hong Kong
Barker D (ed) (1962b) The structure and distribution of muscle receptors. In: Symposium on muscle receptors. Hong Kong University Press, Hong Kong, pp 227–240
Barker D (1966) Three types of motor ending in cat spindles. J Physiol 186:27–28P
Barker D (1974a) The morphology of muscle receptors. In: Hunt CC (ed) Muscle receptors. Springer, Berlin Heidelberg New York, pp 1–190 (Handbook of sensory physiology, vol 3, pt 2)
Barker D (1974b) Symposium on muscle spindles. J Anat 119:193–207
Barker D, Hunt JP (1964) Mammalian intrafusal muscle fibres. Nature 203:1193
Barker D, Ip MC (1961) A study of single and tandem types of muscle spindle in the cat. Proc R Soc Lond (Biol) 154:377–397
Barker D, Ip MC (1965) The motor innervation of cat and rabbit muscle spindles. J Physiol 177:27–28P
Barker D, Milburn A (1982) Development of cat muscle spindles. J Physiol 325:85P
Barker D, Saito M (1981) Autonomic innervation of receptors and muscle fibres in cat skeletal muscle. Proc R Soc Lond (Biol) 212:317–332
Barker D, Stacey MJ (1970) Rabbit intrafusal fibres. J Physiol 210:70–72P
Barker D, Stacey MJ (1981) On the innervation of bag_1 fibres in cat muscle spindles by static γ-axons. J Physiol 320:93P

Barker D, Stacey MJ, Adal MN (1970) Fusimotor innervation in the cat. Philos Trans R Soc Lond (Biol) 258:315–346
Barker D, Bessou P, Jankowska E, Pagès B, Stacey MJ (1972) Distribution des axones fusimoteurs statiques et dynamiques aux fibres musculaires intrafusales chez le chat. C R Hebd Séances Acad Sci (III) 275:2527–2530
Barker D, Emonet-Dénand F, Laporte Y, Proske U, Stacey MJ (1973) Morphological identification and intrafusal distribution of the endings of static fusimotor axons in the cat. J Physiol 230:405–427
Barker D, Banks RW, Harker DW, Milburn A, Stacey MJ (1976a) Studies of the histochemistry, ultrastructure, motor innervation, and regeneration of mammalian intrafusal muscle fibres. Prog Brain Res 44:67–88
Barker D, Emonet-Dénand F, Harker DW, Jami L, Laporte Y (1976b) Distribution of fusimotor axons to intrafusal muscle fibres in cat tenuissimus spindles as determined by the glycogen-depletion method. J Physiol 261:49–69
Barker D, Emonet-Dénand F, Harker DW, Jami L, Laporte Y (1977) Types of intra- and extrafusal muscle fibre innervated by dynamic skeleto-fusimotor axons in cat peroneus brevis and tenuissimus muscles, as determined by the glycogen-depletion method. J Physiol 266:713–726
Barker D, Bessou P, Jankowska E, Pagès B, Stacey MJ (1978) Identification of intrafusal muscle fibres activated by single fusimotor axons and injected with fluorescent dye in cat tenuissimus spindles. J Physiol 275:149–165
Barker D, Emonet-Dénand F, Laporte Y, Stacey MJ (1980) Identification of the intrafusal endings of skeletofusimotor axons in the cat. Brain Res 185:227–237
Baumann TK, Hulliger M (to be published, a) The dependence of the responses of muscle spindle Ia afferents to sinusoidal stretch on the velocity of concomitant triangular movement. J Physiol
Baumann TK, Hulliger M (to be published, b) The power law description of the velocity response of the muscle spindle Ia afferent and its dependence on the parameters of mechanical stimulation. J Neurophysiol
Baumann TK, Emonet-Dénand F, Hulliger M (1982) Long lasting after-effects of fusimotor stimulation on primary spindle afferents, revealed during slow movements. Brain Res 232:460–465
Baumann TK, Emonet-Dénand F, Hulliger M (1983a) Temporal characteristics of the sensitivity-enhancing after-effects of fusimotor activity on spindle Ia afferents. Brain Res 258:139–143
Baumann TK, Emonet-Dénand F, Hulliger M (1983b) After-effects of fusimotor activity: long-lasting enhancement of the dynamic sensitivity of Ia muscle spindle afferents following stimulation of dynamic or static gamma axons. Exp Brain Res (Suppl) 7:266–271
Bergmans J, Grillner S (1968a) Changes in dynamic sensitivity of primary endings of muscle spindle afferents induced by DOPA. Acta Physiol Scand 74:629–636
Bergmans J, Grillner S (1968b) Monosynaptic control of γ-motoneurones from the lower brain stem. Experientia 24:146–147
Bergmans J, Grillner S (1969) Reciprocal control of spontaneous activity and reflex effects in static and dynamic flexor γ-motoneurones revealed by an injection of DOPA. Acta Physiol Scand 77:106–124
Bessou P, Laporte Y (1961) Étude des récepteurs musculaires innervés par les fibres afférentes du groupe III (fibres myélinisées fines), chez le chat. Arch Ital Biol 99: 293–321
Bessou P, Pagès B (1969) Spindle secondary ending responses elicited by stimulation of static fusimotor axons. J Physiol 202:569–584
Bessou P, Pagès B (1972) Intracellular potentials from intrafusal muscle fibres evoked by stimulation of static and dynamic fusimotor axons in the cat. J Physiol 227: 709–727
Bessou P, Pagès B (1975) Cinematographic analysis of contractile events produced in intrafusal muscle fibres by stimulation of static and dynamic fusimotor axons. J Physiol 252:397–427

Bessou P, Emonet-Dénand F, Laporte Y (1963) Occurrence of intrafusal muscle fibre innervation by branches of slow α-motor fibres in the cat. Nature 198:594–595
Bessou P, Emonet-Dénand F, Laporte Y (1965) Motor fibres innervating extrafusal and intrafusal muscle fibres in the cat. J Physiol 180:649–672
Bessou P, Laporte Y, Pagès B (1966) Similitude des effets (statiques ou dynamiques) exercés par des fibres fusimotrices uniques sur les terminaisons primaires de plusieurs fuseaux chez le chat. J Physiol (Paris) 58:31–40
Bessou P, Laporte Y, Pagès B (1968) Frequencygrams of spindle primary endings elicited by stimulation of static and dynamic fusimotor fibres. J Physiol 196:47–63
Bessou P, Joffroy M, Pagès B (1981) Efferents and afferents in an intact muscle nerve: background activity and effects of sural nerve stimulation in the cat. J Physiol 320:81–102
Bessou P, Joffroy M, Montoya R, Pagès B (1984) Effects of triceps stretch by ankle flexion on intact afferents and efferents of gastrocnemius in the decerebrate cat. J Physiol 346:73–91
Bianconi R, van der Meulen JP (1963) The response to vibration of the end organs of mammalian muscle spindles. J Neurophysiol 26:177–190
Binder MD, Stuart DG (1980) Responses of Ia and spindle group II afferents to single motor unit contractions. J Neurophysiol 43:621–629
Binder MD, Kroin JS, Moore GP, Stauffer EK, Stuart DG (1976) Correlation analysis of muscle spindle responses to single motor unit contractions. J Physiol 257:325–336
Bottermann BR, Eldred E (1982) Static stretch sensitivity of Ia and II afferents in the cat's gastrocnemius. Pflügers Arch 395:204–211
Bottermann BR, Binder MD, Stuart DG (1978) Functional anatomy of the association between motor units and muscle receptors. Am Zool 18:135–152
Boyd IA (1956) The tenuissimus muscle of the cat. J Physiol 133:35–36P
Boyd IA (1962) The structure and innervation of the nuclear bag muscle fibre system and the nuclear chain muscle fibre system in mammalian muscle spindles. Philos Trans R Soc Lond (Biol) 245:81–136
Boyd IA (1976a) The response of fast and slow nuclear bag fibres and nuclear chain fibres in isolated cat muscle spindles to fusimotor stimulation, and the effect of intrafusal contraction on the sensory endings. Q J Exp Physiol 61:203–254
Boyd IA (1976b) The mechanical properties of dynamic nuclear bag fibres, static nuclear bag fibres and nuclear chain fibres in isolated cat muscle spindles. Prog Brain Res 44:33–50
Boyd IA (1981a) The muscle spindle controversy. Sci Prog 67:205–221
Boyd IA (1981b) The action of the three types of intrafusal fibre in isolated cat muscle spindles on the dynamic and length sensitivities of primary and secondary sensory endings. In: Taylor A, Prochazka A (eds) Muscle receptors and movement. MacMillan, London, pp 17–32
Boyd IA, Davey MR (1968) Composition of peripheral nerves. Livingstone, Edinburgh
Boyd IA, Ward J (1975) Motor control of nuclear bag and nuclear chain intrafusal fibres in isolated living muscle spindles from the cat. J Physiol 244:83–112
Boyd IA, Ward J (1982) The diagnosis of nuclear chain intrafusal fibre activity from the nature of the group Ia and group II afferent discharge of isolated cat muscle spindles. J Physiol 321:17–18P
Boyd IA, Gladden MH, McWilliam PN, Ward J (1975) "Static" and "dynamic" nuclear bag fibres in isolated cat muscle spindles. J Physiol 250:11–12P
Boyd IA, Gladden MH, McWilliam PN, Ward J (1977) Control of dynamic and static nuclear bag fibres and nuclear chain fibres by gamma and beta axons in isolated cat muscle spindles. J Physiol 265:133–162
Boyd IA, Gladden MH, Ward J (1983a) Suppression of the length sensitivity of primary sensory endings by nuclear chain fibres in cat muscle spindles. J Physiol 339: 59P

Boyd IA, Gladden MH, Ward J (1983b) Two types of static γ-axon having a predominantly static bag_2 fibre action or predominantly chain fibre action in the several cat muscle spindles they supply. J Physiol 343:110P

Brokensha G, Westbury DR (1974) Adaptation of the discharge of frog muscle spindles following a stretch. J Physiol 242:383–403

Brown AG, Fyffe REW (1981) Direct observations on the contacts made between Ia afferent fibres and α-motoneurones in the cat's lumbosacral spinal cord. J Physiol 313:121–140

Brown MC (1971a) A comparison of the spindles in two different muscles of the frog. J Physiol 216:553–563

Brown MC (1971b) The responses of frog muscle spindles and fast and slow muscle fibres to a variety of mechanical inputs. J Physiol 218:1–17

Brown MC, Butler RG (1973) Studies on the site of termination of static and dynamic fusimotor fibres within muscle spindles of the tenuissimus muscle of the cat. J Physiol 233:553–573

Brown MC, Butler RG (1975) An investigation into the site of termination of static gamma fibres within muscle spindles of the cat peroneus longus muscle. J Physiol 247:131–143

Brown MC, Butler RG (1976) Regeneration of afferent and efferent fibres to muscle spindles after nerve injury in adult cats. J Physiol 260:253–266

Brown MC, Matthews PBC (1966) On the subdivision of the efferent fibres to muscle spindles into static and dynamic fusimotor fibres. In: Andrew BL (ed) Control and innervation of skeletal muscle. Truex, Oxford, pp 18–31

Brown MC, Crowe A, Matthews PBC (1965) Observations on the fusimotor fibres of the tibialis posterior muscle of the cat. J Physiol 177:140–159

Brown MC, Engberg I, Matthews PBC (1967a) The relative sensitivity to vibration of muscle receptors of the cat. J Physiol 192:773–800

Brown MC, Engberg I, Matthews PBC (1967b) Fusimotor stimulation and the dynamic sensitivity of the secondary ending of the muscle spindle. J Physiol 189:545–550

Brown MC, Lawrence DG, Matthews PBC (1968a) Reflex inhibition by Ia afferent input of spontaneously discharging motoneurones in the decerebrate cat. J Physiol 198:5–7P

Brown MC, Lawrence DG, Matthews PBC (1968b) Antidromic inhibition of presumed fusimotor neurones by repetitive stimulation of the ventral root of the decerebrate cat. Experientia 24:1210

Brown MC, Goodwin GM, Matthews PBC (1969a) After-effects of fusimotor stimulation on the response of muscle spindle primary endings. J Physiol 205:677–694

Brown MC, Lawrence DG, Matthews PBC (1969b) Static fusimotor fibres and the position sensitivity of muscle spindle receptors. Brain Res 14:173–187

Browne JS (1975) The responses of muscle spindles in sheep extraocular muscles. J Physiol 251:483–496

Buchthal F, Schmalbruch H (1980) Motor unit of mammalian muscle. Physiol Rev 60: 90–142

Burg D, Szumski AJ, Struppler A, Velho F (1973) Afferent and efferent activation of human muscle receptors involved in reflex and voluntary contraction. Exp Neurol 41:754–768

Burg D, Szumski AJ, Struppler A, Velho F (1974) Assessment of fusimotor contribution to reflex reinforcement in humans. J Neurol Neurosurg Psychiatry 37:1012–1021

Burg D, Szumski AJ, Struppler A, Velho F (1976) Influence of a voluntary innervation on human muscle spindle sensitivity. In: Shahani M (ed) The motor system: Neurophysiology and muscle mechanisms, vol 4. Elsevier, Amsterdam, pp 95–110

Burgess PR, Clark FJ (1969) Characteristics of knee joint receptors in the cat. J Physiol 203:317–335

Burke D (1980) A reassessment of the muscle spindle contribution to muscle tone in normal and spastic man. In: Feldman RG, Young RR, Koella WP (eds) Spasticity: Disordered motor control. Symposia Specialists, Miami, pp 260–286

Burke D (1981) The activity of human muscle spindle endings in normal motor behavior. Int Rev Physiol 20:91–136
Burke D, Eklund G (1977) Muscle spindle activity in man during standing. Acta Physiol Scand 100:187–199
Burke D, Hagbarth K-E, Löfstedt L, Wallin BG (1976a) The responses of human muscle spindle endings to vibration of non-contracting muscles. J Physiol 261:673–693
Burke D, Hagbarth K-E, Löfstedt L, Wallin BG (1976b) The response of human muscle spindle endings to vibration during isometric contraction. J Physiol 261:695–711
Burke D, Hagbarth K-E, Löfstedt L (1978a) Muscle spindle responses in man to changes in load during accurate position maintenance. J Physiol 276:159–164
Burke D, Hagbarth K-E, Löfstedt L (1978b) Muscle spindle activity in man during shortening and lengthening contractions. J Physiol 277:131–142
Burke D, Hagbarth K-E, Skuse NF (1978c) Recruitment order of human spindle endings in isometric voluntary contractions. J Physiol 285:101–112
Burke D, Hagbarth K-E, Skuse NF (1979a) Voluntary activation of spindle endings in human muscles temporarily paralysed by nerve pressure. J Physiol 287:329–336
Burke D, Skuse NF, Stuart DG (1979b) The regularity of muscle spindle discharge in man. J Physiol 291:277–290
Burke D, McKeon B, Skuse NF, Westerman RA (1980a) Anticipation and fusimotor activity in preparation for a voluntary contraction. J Physiol 306:337–348
Burke D, McKeon B, Westerman RA (1980b) Induced changes in the threshold for voluntary activation of human spindle endings. J Physiol 302:171–181
Burke RE (1981) Motor units: anatomy, physiology, and functional organization. In: Brookhart JM, Mountcastle VB, Brooks VB (eds) Motor control, part 1. American Physiological Society, Bethesda, pp 345–422 (Handbook of physiology, sect 1, vol 2)
Burke RE, Edgerton VR (1975) Motor unit properties and selective involvement in movement. In: Wilmore JH, Keogh JF (eds) Exercise and sports science reviews, vol 3. Academic, New York, pp 31–69
Butler RG (1980) The organization of muscle spindles in the tenuissimus muscle of the cat during late development. Dev Biol 77:191–212
Cabelguen JM (1981) Static and dynamic fusimotor controls in various hindlimb muscles during locomotor activity in the decorticate cat. Brain Res 213:83–97
Cameron WE, Binder MD, Botterman BR, Reinking RM, Stuart DG (1981) "Sensory partitioning" of cat medial gastrocnemius muscle by its muscle spindles and tendon organs. J Neurophysiol 46:32–47
Celio MR, Lutz H, Jenny E (1981) Myosin isoenzymes in rat muscle spindles. Neurosci Lett (Suppl) 7:100
Chen WJ, Poppele RE (1973) Static fusimotor effect on the sensitivity of mammalian muscle spindles. Brain Res 57:244–247
Chen WJ, Poppele RE (1978) Small-signal analysis of response of mammalian muscle spindles with fusimotor stimulation and a comparison with large-signal responses. J Neurophysiol 41:15–27
Cheney PD, Preston JB (1976a) Classification and response characteristics of muscle spindle afferents in the primate. J Neurophysiol 39:1–8
Cheney PD, Preston JB (1976b) Classification of fusimotor fibres in the primate. J Neurophysiol 39:9–19
Cheney PD, Preston JB (1976c) Effects of fusimotor stimulation on dynamic and position sensitivities of spindle afferents in the primate. J Neurophysiol 39:20–30
Clough JHM, Phillips CG, Sheridan JP (1971) The short latency projection from the baboon's motor cortex to fusimotor neurones of the forearm and hand. J Physiol 216:257–279
Cody FWJ, Lee RWH, Taylor A (1972) A functional analysis of the components of the mesencephalic nucleus of the fifth nerve in the cat. J Physiol 226:249–261
Cody FWJ, Harrison LM, Taylor A (1975) Analysis of activity of muscle spindles of the jaw-closing muscles during normal movements in the cat. J Physiol 253:565–582

Cooper S (1961) The responses of primary and secondary endings of muscle spindles with intact motor innervation during applied stretch. Q J Exp Physiol 46:389–398
Cooper S, Daniel PM (1956) Human muscle spindles. J Physiol 133:1–3P
Cooper S, Daniel PM (1963) Muscle spindles in man: their morphology in the lumbricals and the deep muscles of the neck. Brain 86:563–586
Cooper S, Gladden MH (1974) Elastic fibres and reticulin of mammalian muscle spindles and their functional significance. Q J Exp Physiol 59:367–385
Corbin KB (1940) Observations on the peripheral distribution of the fibres arising in the mesencephalic nucleus of the fifth cranial nerve. J Comp Neurol 73:153–177
Corda M, von Euler C, Lennerstrand G (1965) Proprioceptive innervation of the diaphragm. J Physiol 178:161–177
Corda M, von Euler C, Lennerstrand G (1966) Reflex and cerebellar influences on α- and on "rhythmic" and "tonic" γ-activity in the intercostal muscle. J Physiol 184: 898–923
Critchlow V, von Euler C (1963) Intercostal muscle spindle activity and its gamma motor control. J Physiol 168:820–847
Crowe A, Matthews PBC (1964a) The effects of stimulation of static and dynamic fusimotor fibres on the response to stretching of the primary endings of muscle spindles. J Physiol 174:109–131
Crowe A, Matthews PBC (1964b) Further studies of static and dynamic fusimotor fibres. J Physiol 174:132–151
Cullheim S, Ulfhake B (1979) Observations on the morphology of intracellularly stained γ-motoneurones in relation to their axon conduction velocity. Neurosci Lett 13:47–50
Cussons PD, Hulliger M, Matthews PBC (1977) Effects of fusimotor stimulation on the response of the secondary ending of the muscle spindle to sinusoidal stretching. J Physiol 270:835–850
Davey MR, Taylor A (1966) Activity of jaw muscle stretch receptors recorded from their cell bodies in the midbrain during spontaneous jaw movements. J Physiol 185: 62–63P
Decorte L, Emonet-Dénand F, Harker DW, Jami L, Laporte Y (1984) Glycogen depletion elicited in tenuissimus intrafusal muscle fibres by stimulation of static γ-axons in the cat. J Physiol 346:341–352
Diete-Spiff K, Dodsworth H, Pascoe JE (1962) An analysis of the effect of ether and ethyl chloride on the discharge frequency of gastrocnemius fusimotor neurones in the rabbit. In: Barker D (ed) Symposium on muscle receptors. Hong Kong University Press, Hong Kong, pp 43–47
Durkovic RG (1976) Aftereffects of static or dynamic fusimotor activation on primary afferent discharge. Exp Neurol 50:99–112
Dutia MB (1980) Activation of cat muscle spindle primary, secondary and intermediate sensory endings by suxamethonium. J Physiol 304:315–330
Eagles JP, Purple RL (1974) Afferent fibres with multiple encoder sites. Brain Res 77: 187–193
Eccles JC, Fatt P, Koketsu K (1954) Cholinergic and inhibitory synapses in a pathway from motor axon collaterals to motoneurones. J Physiol 126:524–562
Eccles JC, Eccles RM. Lundberg A (1957a) Synaptic actions on motoneurones in relation to the two components of the group I muscle afferent volley. J Physiol 136: 527–546
Eccles JC, Eccles RM, Lundberg A (1957b) The convergence of monosynaptic excitatory afferents onto many different species of alpha motoneurones. J Physiol 137: 22–50
Eccles JC, Eccles RM, Iggo A, Lundberg A (1960) Electrophysiological studies on gamma motoneurones. Acta Physiol Scand 50:32–40
Eccles RM, Lundberg A (1959) Synaptic actions in motoneurones by afferents which may evoke the flexion reflex. Arch Ital Biol 97:199–221

Edström L, Kugelberg E (1968) Histochemical composition, distribution of fibres and fatiguability of single motor units. J Neurol Neurosurg Psychiatry 31:424–433
Eklund G, von Euler C, Rutkowski S (1964) Spontaneous and reflex activity of intercostal gamma motoneurones. J Physiol 171:139–163
Eldred E, Hagbarth K-E (1954) Facilitation and inhibition of gamma efferents by stimulation of certain skin areas. J Neurophysiol 17:59–65
Eldred E, Granit R, Merton PA (1953) Supraspinal control of the muscle spindles and its significance. J Physiol 122:498–523
Ellaway PH (1968) Antidromic inhibition of fusimotor neurones. J Physiol 198:39–40P
Ellaway PH (1971) Recurrent inhibition of fusimotor neurones exhibiting background discharges in the decerebrate and the spinal cat. J Physiol 216:419–439
Ellaway PH (1972) The variability in discharge of fusimotor neurones in the decerebrate cat. Exp Brain Res 14:105–117;
Ellaway PH (1976) Recurrent inhibition of gamma motoneurones. In: Shahani M (ed) The motor system: neurophysiology and muscle mechanisms, vol 4. Elsevier, Amsterdam, pp 119–126
Ellaway PH, Murphy PR (1980) Autogenetic effects of muscle contraction on extensor gamma motoneurones in the cat. Exp Brain Res 38:305–312
Ellaway PH, Murphy PR (1981) A comparison of the recurrent inhibition of α- and γ-motoneurones in the cat. J Physiol 315:43–58
Ellaway PH, Trott J (1976) Reflex connections from muscle stretch receptors to their own fusimotor neurons. Prog Brain Res 44:113–122
Ellaway PH, Trott JR (1978) Autogenetic reflex action onto gamma motoneurones by stretch of triceps surae in the decerebrate cat. J Physiol 276:49–66
Ellaway PH, Emonet-Dénand F, Joffroy M, Laporte Y (1972) Lack of exclusively fusimotor α-axons in flexor and extensor leg muscles of the cat. J Neurophysiol 35: 149–153
Ellaway PH, Murphy PR, Trott JR (1979) Inhibition of gamma motoneurone discharge by contraction of the homonymous muscle in the decerebrated cat. J Physiol 291:425–441
Ellaway PH, Murphy PR, Tripathi A (1982) Closely coupled excitation of γ-motoneurones by group III muscle afferents with low mechanical threshold in the cat. J Physiol 331:481–498
Emonet-Dénand F, Laporte Y (1969) Frequencygrams of rabbit spindle primary endings elicited by stimulation of fusimotor fibres. J Physiol 201:673–684
Emonet-Dénand F, Laporte Y (1974) Blocage neuromusculaire sélectif des jonctions extrafusales des axones squeletto-fusimoteurs produit par leur stimulation répétitive à fréquence élevée. C R Hebd Séances Acad Sci (III) 279:2083–2085
Emonet-Dénand F, Laporte Y (1975) Proportion of muscle spindles supplied by skeletofusimotor axons (β-axons) in peroneus brevis muscle of the cat. J Neurophysiol 38:1390–1394
Emonet-Dénand F, Laporte Y (1981) Muscle stretch as a way of detecting brief activation of bag_1 fibres by dynamic axons. In: Taylor A, Prochazka A (eds) Muscle receptors and movement. MacMillan, London, pp 67–76
Emonet-Dénand F, Laporte Y (1983) Observations on the effects on spindle primary endings of the stimulation at low frequency of dynamic β-axons. Brain Res 258: 101–104
Emonet-Dénand F, Laporte Y, Pagès B (1966) Fibres fusimotrices statiques et fibres fusimotrices dynamiques chez le lapin. Arch Ital Biol 104:195–213
Emonet-Dénand F, Jankowska E, Laporte Y (1970) Skeleto-fusimotor fibres in the rabbit. J Physiol 210:669–680
Emonet-Dénand F, Joffroy M, Laporte Y (1972) Fibres fusimotrices dont l'action sur la sensibilité phasique des terminaisons primaires dépend de leur fréquence de stimulation. C R Hebd Séances Acad Sci (III) 275:89–91

Emonet-Dénand F, Jami L, Laporte Y (1975) Skeleto-fusimotor axons in hind-limb muscles of the cat. J Physiol 249:153–166
Emonet-Dénand F, Hulliger M, Matthews PBC, Petit J (1977a) Factors affecting modulation in post-stimulus histograms on static fusimotor stimulation. Brain Res 134: 180–184
Emonet-Dénand F, Laporte Y, Matthews PBC, Petit J (1977b) On the subdivision of static and dynamic fusimotor actions on the primary ending of the cat muscle spindle. J Physiol 268:827–861
Emonet-Dénand F, Jami L, Laporte Y (1980a) Histophysiological observations on the skeleto-fusimotor innervation of mammalian spindles. Prog. Clin Neurophysiol 8: 1–11
Emonet-Dénand F, Jami L, Laporte Y, Tankov N (1980b) Glycogen depletion of bag_1 fibres elicited by stimulation of static γ-axons in cat peroneus brevis muscle spindles. J Physiol 302:311–321
Emonet-Dénand F, Laporte Y, Tristant A (1980c) Effects of slow muscle stretch on the responses of primary and secondary endings to small amplitude periodic stretches in de-efferented soleus muscle spindles. Brain Res 191:551–554
Evans MH (1963) Alterations in activity of gamma efferents during distension of the bladder of the cat. J Physiol 165:358–367
Eyzaguirre C (1957) Functional organization of neuromuscular spindle in toad. J Neurophysiol 20:523–542
Eyzaguirre C (1958) Modulation of sensory discharges by efferent spindle excitation. J Neurophysiol 21:465–480
Fidone SJ, Preston JB (1969) Patterns of motor cortex control of flexor and extensor cat fusimotor neurons. J Neurophysiol 32:103–115
Filippi GM, Grassi C, Passatore M (1983) The possible action of the cervical sympathetic nerves on intrafusal fibres of jaw muscle spindles in the rabbit. J Physiol 340:27P
Fromm C, Noth J (1976) Reflex responses of gamma motoneurones to vibration of the muscle they innervate. J Physiol 256:117–136
Fromm C, Haase J, Noth J (1974) Length-dependent autogenetic inhibition of extensor γ-motoneurones in the decerebrate cat. Pflügers Arch 346:251–262
Fromm C, Noth J, Thilmann A (1976) Inhibition of extensor γ-motoneurones by antagonistic primary and secondary spindle afferents. Pflügers Arch 363:81–86
Fulton JF, Pi-Suñer J (1928) A note concerning the probable function of various afferent end-organs in skeletal muscle. Am J Physiol 83:554–562
Gilman S (1968) A crossed cerebellar influence on muscle spindle primaries. Brain Res 8:216–219
Gilman S (1969) The mechanisms of cerebellar hypotonia. An experimental study in the monkey. Brain 92:621–638
Gilman S, McDonald WI (1967) Cerebellar facilitation of muscle spindle activity. J Neurophysiol 30:1494–1512
Gilman S, Lieberman JS, Copack P (1971a) A thalamic mechanism of postural control. Int J Neurol 8:260–275
Gilman S, Marco LA, Ebel HC (1971b) Effects of medullary pyramidotomy in the monkey. II. Abnormalities of spindle afferent responses. Brain 94:515–530
Gladden MH (1976) Structural features relative to the function of intrafusal muscle fibres in the cat. Prog Brain Res 44:51–59
Gladden MH (1981) The activity of intrafusal muscle fibres during central stimulation in the cat. In: Taylor A, Prochazka A (eds) Muscle receptors and movement. MacMillan, London, pp 109–122
Gladden MH, McWilliam PN (1977) The activity of intrafusal fibres during cortical stimulation in the cat. J Physiol 273:28–29P
Goodwin GM, Luschei ES (1974) Effects of destroying spindle afferents from jaw muscles on mastication in monkeys. J Neurophysiol 37:967–981
Goodwin GM, Luschei ES (1975) Discharge of spindle afferents from jaw-closing muscles during chewing in alert monkeys. J Neurophysiol 38:560–571

Goodwin GM, McCloskey DI, Matthews PBC (1972) The contribution of muscle afferents to kinaesthesia shown by vibration-induced illusions of movement and by the effects of paralysing joint afferents. Brain 95:705–748
Goodwin GM, Hulliger M, Matthews PBC (1975) The effects of fusimotor stimulation during small amplitude stretching on the frequency-response of the primary ending of the mammalian muscle spindle. J Physiol 253:175–206
Goodwin GM, Hulliger M, Matthews PBC (1976) Studies on muscle spindle primary endings with sinusoidal stretching. Prog Brain Res 44:89–98
Goslow GE, Reinking RM, Stuart DG (1973) The cat step cycle. Hindlimb joint angles and muscle length during unrestrained locomotion. J Morphol 141:1–41
Gottlieb S, Taylor A (1983) Interpretation of fusimotor activity in cat masseter nerve during reflex jaw movements. J Physiol 345:423–438
Granit R (1955) Receptors and sensory perception. Yale University Press, New Haven
Granit R (1970) The basis of motor control. Academic Press, London
Granit R (1979) Interpretation of supraspinal effects on the gamma system. Prog Brain Res 50:147–154
Granit R, Holmgren B (1955) Two pathways from brain stem to gamma ventral horn cells. Acta Physiol Scand 35:93–108
Granit R, Kaada BR (1952) Influence of stimulation of central nervous structures on muscle spindles in cat. Acta Physiol Scand 27:130–160
Granit R, Job C, Kaada BR (1952) Activation of muscle spindles in pinna reflex. Acta Physiol Scand 27:161–168
Granit R, Holmgren B, Merton PA (1955) The two routes for excitation of muscle and their subservience to the cerebellum. J Physiol 130:213–224
Granit R, Pascoe JE, Steg G (1957) The behaviour of tonic α- and γ-motoneurones during stimulation of recurrent collaterals. J Physiol 138:381–400
Gray EG (1957) The spindle and extrafusal innervation of a frog muscle. Proc R Soc Lond (Biol) 146:416–430
Gregory JE, Prochazka A, Proske U (1977) Responses of muscle spindles to stretch after a period of fusimotor activity compared in freely moving and anaesthetized cats. Neurosci Lett 4:67–72
Grigg P (1975) Mechanical factors influencing response of joint afferent neurons from cat knee. J Neurophysiol 38:1473–1484
Grigg P, Greenspan BJ (1977) Response of primate joint afferent neurons to mechanical stimulation of knee joint. J Neurophysiol 40:1–8
Grigg P, Preston JB (1971) Baboon flexor and extensor fusimotor neurons and their modulation by motor cortex. J Neurophysiol 34:428–436
Grillner S (1969a) The influence of DOPA on the static and dynamic fusimotor activity to the triceps surae of the spinal cat. Acta Physiol Scand 77:490–509
Grillner S (1969b) Supraspinal and segmental control of static and dynamic γ-motoneurones in the cat. Acta Physiol Scand (Suppl) 327:1–34
Grillner S (1976) Some aspects on the descending control of the spinal circuits generating locomotor movements. In: Herman RM, Grillner S, Stein PSG, Stuart DG (eds) Neural control of locomotion. Plenum, New York, pp 351–375
Grillner S (1981) Control of locomotion in bipeds, tetrapods, and fish. In: Brookhart JM, Mountcastle VB, Brooks VB (eds) Motor control, part 2. American Physiological Society, Bethesda, pp 1179–1236 (Handbook of physiology, sect 1, vol 2)
Grillner S, Hongo T, Lundberg A (1967) The effect of DOPA on the spinal cord. 7. Reflex activation of static γ-motoneurones from the flexor reflex afferents. Acta Physiol Scand 70:403–411
Grillner S, Hongo T, Lund S (1969) Descending monosynaptic and reflex control of γ-motoneurones. Acta Physiol Scand 75:592–613
Hagbarth K-E (1952) Excitatory and inhibitory skin areas for flexor and extensor motoneurones. Acta Physiol Scand (Suppl 94) 26:1–58
Hagbarth K-E (1979) Exteroceptive, proprioceptive, and sympathetic activity recorded with microelectrodes from human peripheral nerves. Mayo Clin Proc 54:353–365

Hagbarth K-E (1981) Fusimotor and stretch reflex functions studied in recordings from muscle spindle afferents in man. In: Taylor A, Prochazka A (eds) Muscle receptors and movement. MacMillan, London, pp 277–286
Hagbarth K-E, Vallbo ÅB (1968) Discharge characteristics of human muscle afferents during muscle stretch and contraction. Exp Neurol 22:674–694
Hagbarth K-E, Young RR (1979) Participation of the stretch reflex in human physiological tremor. Brain 102:509–526
Hagbarth K-E, Wallin G, Burke D, Löfstedt L (1975a) Effects of the Jendrassik manoeuvre on muscle spindle activity in man. J Neurol Neurosurg Psysiatry 38:1143–1153
Hagbarth K-E, Wallin G, Löfstedt L (1975b) Muscle spindle activity in man during voluntary fast alternating movements. J Neurol Neurosurg Psychiatry 38:625–635
Hagbarth K-E, Wallin G, Löfstedt L, Aquilonius S-M (1975c) Muscle spindle activity in alternating tremor of parkinsonism and in clonus. J Neurol Neurosurg Psychiatry 38:636–641
Hagbarth K-E, Hägglund JV, Wallin EU, Young RR (1981) Grouped spindle and electromyographic responses to abrupt wrist extension movements in man. J Physiol 312:81–96
Harker DW, Jami L, Laporte Y, Petit J (1977) Fast-conducting skeletofusimotor axons supplying intrafusal chain fibres in the cat peroneus tertius muscle. J Neurophysiol 40:791–799
Harriman DGF, Parker PL, Elliot BJ (1974) The histochemistry of human intrafusal muscle fibres. J Anat 119:205–206
Harvey RJ, Matthews PBC (1961) The response of deefferented muscle spindle endings in the cat's soleus to slow extension of the muscle. J Physiol 157:170–192
Hasan Z (1983) A model of spindle afferent response to muscle stretch. J Neurophysiol 49:989–1006
Hasan Z, Houk JC (1975a) Analysis of response properties of deefferented mammalian spindle receptors based on frequency response. J Neurophysiol 38:663–672
Hasan Z, Houk JC (1975b) Transition in sensitivity of spindle receptors that occurs when muscle is stretched more than a fraction of a millimeter. J Neurophysiol 38: 673–689
Henatsch HD, Manni E, Wilson JH, Dow RS (1964) Linked and independent responses of tonic α and γ hind-limb motoneurons to deep cerebellar stimulation. J Neurophysiol 27:172–192
Henneman E, Mendell LM (1981) Functional organization of motoneuron pool and its inputs. In: Brookhart JM, Mountcastle VB, Brooks VB (eds) Motor control, part 1. American Physiological Society, Bethesda, pp 423–507 (Handbook of physiology, sect 1, vol 2)
Henneman E, Somjen G, Carpenter DO (1965a) Functional significance of cell size in spinal motoneurons. J Neurophysiol 28:560–580
Henneman E, Somjen G, Carpenter DO (1965b) Excitability and inhibitibility of motoneurons of different sizes. J Neurophysiol 28:599–620
Hill DK (1968) Tension due to interaction between the sliding filaments in resting striated muscle. J Physiol 199:637–684
Hnik P, Kucera J, Kidd GL (1970) Increased sensory outflow from muscles following tetanic stimulation of alpha motor fibres. Physiol Bohemoslov 19:49–54
Hoffer JA, Loeb GE (1983) A technique for reversible fusimotor blockade during chronic recording from spindle afferents in walking cats. Exp Brain Res (Suppl) 7:272–279
Hoffer JA, Loeb GE, Pratt CA (1981a) Single unit conduction velocities from averaged nerve cuff electrode records in freely moving cats. J Neurosci Methods 4:211–225
Hoffer JA, O'Donovan MJ, Pratt CA, Loeb GE (1981b) Discharge patterns of hindlimb motoneurons during normal cat locomotion. Science 213:466–468

Holm W, Padeken D, Schäfer SS (1981) Characteristic curves of the dynamic response of primary muscle spindle endings with and without gamma stimulation. Pflügers Arch 391:163–170
Holmqvist B, Lundberg A (1961) Differential supraspinal control of synaptic actions evoked by volleys in the flexion reflex afferents in alpha motoneurones. Acta Physiol Scand (Suppl 186) 54:1–51
Homma S (ed) (1976) Understanding the stretch reflex. Prog Brain Res 44
Hongo T, Jankowska E, Lundberg A (1969) The rubrospinal tract. I. Effects on alpha motoneurones innervating hindlimb muscles of cats. Exp Brain Res 7:344–364
Houk JC (1972) The phylogeny of muscular control configurations. In: Drischel H, Dettmar P (eds) Biocybernetics, vol IV. Jena: Fischer, pp 125–144
Houk JC, Rymer WZ, Crago PE (1981) Dependence of dynamic response of spindle receptors on muscle length and velocity. J Neurophysiol 46:143–166
Hulliger M (1976) Some quantitative studies on muscle receptors. Doctoral thesis, Oxford University
Hulliger M (1979) The responses of primary spindle afferents to fusimotor stimulation at constant and abruptly changing rates. J Physiol 294:461–482
Hulliger M (1981) Muscle spindle afferent units. Functional properties with possible significance in spasticity. In: Bauer HJ, Koella WP, Struppler A (eds) Therapie der Spastik. Verlag für angewandte Wissenschaften, München, pp 55–70
Hulliger M, Noth J (1979) Static and dynamic fusimotor interaction and the possibility of multiple pace-makers operating in the cat muscle spindle. Brain Res 173:21–28
Hulliger M, Prochazka A (1983) A new simulation method to deduce fusimotor activity from afferent discharge recorded in freely moving cats. J Neurosci Methods 8: 197–204
Hulliger M, Vallbo ÅB (1979a) The responses of muscle spindle afferents during voluntary tracking movements. Load-dependent servo assistance? Brain Res 166:401–404
Hulliger M, Vallbo ÅB (1979b) The responses of muscle spindle afferents to small irregularities of voluntary movement in man. Neurosci Lett (Suppl) 3:98
Hulliger M, Matthews PBC, Noth J (1977a) Static and dynamic fusimotor action on the response of Ia fibres to low-frequency sinusoidal stretching of widely ranging amplitude. J Physiol 267:811–838
Hulliger M, Matthews PBC, Noth J (1977b) Effects of combining static and dynamic fusimotor stimulation on the response of the muscle spindle primary ending to sinusoidal stretching. J Physiol 267:839–856
Hulliger M, Matthews PBC, Noth J (1977c) Alternation between occlusion and summation during combined fusimotor excitation of Ia afferents. J Physiol 268:27–28P
Hulliger M, Nordh E, Thelin AE, Vallbo ÅB (1979) The responses of afferent fibres from the glabrous skin of the hand during voluntary finger movements in man. J Physiol 291:233–249
Hulliger M, Nordh E, Vallbo ÅB (1982) The absence of position response in spindle afferent units from human finger muscles during accurate position holding. J Physiol 322:167–179
Hulliger M, Baumann TK, Emonet-Dénand F (1983) Static γ-motoneurones activated at low stimulation rates and short muscle lengths may enhance Ia sensitivity to sinusoidal stretch. Proc Int Union Physiol Sci 15:86
Hulliger M, Zangger P, Prochazka A, Appenteng K (1984) Fusimotor "Set" vs α-γ linkage in voluntary movement in cat. In: Struppler A, Weindl A (eds) Proceedings of the 7th international congress of electromyography. Springer, Berlin Heidelberg New York, pp 57–64
Hultborn H (1976) Transmission in the pathway of reciprocal Ia inhibition to motoneurones and its control during the tonic stretch reflex. Prog Brain Res 44:235–255
Hultborn H, Lindström S, Wigström H (1979) On the function of recurrent inhibition in the spinal cord. Exp Brain Res 37:399–403
Hunt CC (1951) The reflex activity of mammalian small nerve fibres. J Physiol 115: 456–469

Hunt CC (1954) Relation of function to diameter in afferent fibres of muscle nerves. J Gen Physiol 38:117–131
Hunt CC (1974) The physiology of muscle receptors. In: Hunt CC (ed) Muscle receptors. Springer, Berlin Heidelberg New York, pp 191–234 (Handbook of sensory physiology, vol 3, pt 2)
Hunt CC, Kuffler SW (1951a) Further study of efferent small-nerve fibres to mammalian muscle spindles. Multiple spindle innervation and activity during contraction. J Physiol 113:283–297
Hunt CC, Kuffler SW (1951b) Stretch receptor discharges during muscle contraction. J Physiol 113:298–315
Hunt CC, Ottoson D (1976) Initial burst of primary endings of isolated mammalian muscle spindles. J Neurophysiol 39:324–330
Hunt CC, Ottoson D (1977) Responses of primary and secondary endings of isolated mammalian muscle spindles to sinusoidal length changes. J Neurophysiol 40:1113–1120
Hunt CC, Paintal AS (1958) Spinal reflex regulation of fusimotor neurones. J Physiol 143:195–212
Hunt CC, Perl ER (1960) Spinal reflex mechanisms concerned with skeletal muscle. Physiol Rev 40:538–579
Hunt CC, Wilkinson RS (1980) An analysis of receptor potential and tension of isolated cat muscle spindles in response to sinusoidal stretch. J Physiol 302:241–262
Hunt CC, Jami L, Laporte Y (1982) Effects of stimulating the lumbar sympathetic trunk on cat hindlimb muscle spindles. Arch Ital Biol 120:371–384
Hutton RS, Smith JL, Eldred E (1973) Postcontraction sensory discharge and its source. J Neurophysiol 36:1090–1103
Hutton RS, Smith JL, Eldred E (1975) Persisting changes in sensory and motor activity of a muscle following its reflex activation. Pflügers Arch 353:327–336
Hyde D, Scott JJA (1983) Responses of cat peroneus brevis muscle spindle afferents during recovery from nerve-crush injury. J Neurophysiol 50:344–357
Iggo A (1961) Non-myelinated afferent fibres from mammalian skeletal muscle. J Physiol 155:52–53P
Inoue H, Morimoto T, Kawamura Y (1981) Response characteristics and classification of muscle spindles of the masseter muscle in the cat. Exp Neurol 74:548–560
Jami L, Petit J (1978) Fusimotor actions on sensitivity of spindle secondary endings to slow muscle stretch in cat peroneus tertius. J Neurophysiol 41:860–869
Jami L, Petit J (1979) Dynamic and static responses of primary and secondary spindle endings of the cat peroneus tertius muscle. J Physiol 296:109P
Jami L, Petit J (1981) Fusimotor actions on the sensitivity of spindle secondary endings. In: Taylor A, Prochazka A (eds) Muscle receptors and movement. MacMillan, London, pp 51–66
Jami L, Lan-Couton D, Malmgren K, Petit J (1978) "Fast" and "slow" skeleto-fusimotor innervation in cat tenuissimus spindles: a study with the glycogen-depletion method. Acta Physiol Scand 103:284–298
Jami L, Lan-Couton D, Malmgren K, Petit J (1979) Histophysiological observations on fast skeleto-fusimotor axons. Brain Res 164:53–59
Jami L, Murthy KSK, Petit J (1982) A quantitative study of skeletofusimotor innervation in the cat peroneus tertius muscle. J Physiol 325:125–144
Jankowska E (1979) New observations on neuronal organization of reflexes from tendon organ afferents and their relation to reflexes evoked from muscle spindle afferents. Prog Brain Res 50:29–36
Jankowska E, Jukes MGM, Lund S, Lundberg A (1967a) The effect of DOPA on the spinal cord. 5. Reciprocal organization of pathways transmitting excitatory action to alpha motoneurones of flexors and extensors. Acta Physiol Scand 70:369–388
Jankowska E, Jukes MGM, Lund S, Lundberg A (1967b) The effect of DOPA on the spinal cord. 6. Half-centre organization of interneurones transmitting effects from the flexor reflex afferents. Acta Physiol Scand 70:389–402

Jansen JKS, Matthews PBC (1962) The central control of the dynamic response of muscle spindle receptors. J Physiol 161:357–378
Jerge CR (1963) Organization and function of the trigeminal mesencephalic nucleus. J Neurophysiol 26:379–392
Johansson H (1981) Reflex control of γ-motoneurones. Umeå University medical dissertations, new series no 70
Kato M, Fukushima K (1974) Effect of differential blocking of motor axons on antidromic activation of Renshaw cells in the cat. Exp Brain Res 20:135–143
Kato M, Takamura H, Fujimori B (1964) Studies on effects of pyramid stimulation upon flexor and extensor motoneurones and gamma motoneurones. Jpn J Physiol 14:34–44
Katz B (1949) Efferent regulation of muscle spindle in the frog. J Exp Biol 26:201–217
Kennedy WR (1970) Innervation of normal human muscle spindles. Neurology (Minneap) 20:463–475
Kennedy WR, Poppele RE, Quick DC (1980) Mammalian muscle spindles. In: Sumner AJ (ed) The physiology of peripheral nerve disease. Saunders, Philadelphia, pp 74–132
Kidd GL (1964) A persistent excitation of muscle-spindle receptor endings in the rat following ventral root filament stimulation. J Physiol 170:39–52
Kidd GL, Kucera J, Vaillant CH (1971) The influence of the interstitial concentration of K^+ on the activity of muscle receptors. Physiol Bohemoslov 20:95–108
Kirkwood PA, Sears TA (1974) Monosynaptic excitation of motoneurones from secondary endings of muscle spindles. Nature 252:243–244
Kirkwood PA, Sears TA (1975) Monosynaptic excitation of motoneurones from muscle spindle secondary endings of intercostal and triceps surae muscles in the cat. J Physiol 245:64–66P
Kniffki KD, Mense S, Schmidt RF (1978) Responses of group IV afferent units from skeletal muscle to stretch, contraction and chemical stimulation. Exp Brain Res 31: 511–522
Kniffki KD, Schomburg ED, Steffens H (1981) Effects from fine muscle and cutaneous afferents on spinal locomotion in cats. J Physiol 319:543–554
Kobayashi Y, Oshima K, Tasaki I (1952) Analysis of afferent and efferent systems in the muscle nerve of the toad and cat. J Physiol 117:152–171
Koeze TH (1968) The response to stretch of muscle spindle afferents of baboon's tibialis anticus and the effect of fusimotor stimulation. J Physiol 197:107–121
Koeze TH (1973a) Muscle spindle afferent studies in the baboon. J Physiol 229:297–317
Koeze TH (1973b) Thresholds of cortical activation of baboon α- and γ-motoneurones during halothane anaesthesia. J Physiol 229:319–337
Kornhauser D, Bromberg MB, Gilman S (1982) Effects of lesions of fastigial nucleus on static and dynamic responses of muscle spindle primary afferents in the cat. J Neurophysiol 47:977–986
Kozeka K, Ontell M (1981) The three-dimensional cytoarchitecture of developing murine muscle spindles. Dev Biol 87:133–147
Kubota K, Masegi T (1977) Muscle spindle supply to the human jaw muscle. J Dent Res 56:901–910
Kucera J (1977) Histochemistry of intrafusal muscle fibers outside the spindle capsule. Am J Anat 148:427–432
Kucera J (1980a) Motor innervation of the cat muscle spindle studied by the cholinesterase technique. Histochemistry 67:291–309
Kucera J (1980b) Histochemical study of long nuclear chain fibers in the cat muscle spindle. Anat Rec 198:567–580
Kucera J (1980c) Motor nerve terminals of cat nuclear chain fibers studied by the cholinesterase technique. Neuroscience 5:403–411

Kucera J (1981a) The occurrence of "mixed" nuclear bag intrafusal fibers in the cat muscle spindle. Histochemistry 72:123–131

Kucera J (1981b) Histochemical profiles of cat intrafusal muscle fibers and their motor innervation. Histochemistry 73:397–418

Kucera J (1982a) Morphometric studies on tenuissimus muscle spindles in the cat. J Morphol 171:137–150

Kucera J (1982b) A study of motor nerve terminals on cat nuclear bag_1 intrafusal muscle fibers using the ChE staining technique. Anat Rec 202:407–418

Kucera J (1982c) The topography of long nuclear chain intrafusal fibers in the cat muscle spindle. Histochemistry 74:183–197

Kucera J (1982d) Appearance of sensory nerve terminals in cat muscle spindles stained for NADH-tetrazolium reductase. Histochemistry 74:493–503

Kucera J (1982e) A study of sensory innervation to long nuclear chain intrafusal fibers in the cat muscle spindle. Histochemistry 75:113–121

Kucera J (1982f) One-bag-fiber muscle spindles in tenuissimus muscles of the cat. Histochemistry 76:315–328

Kucera J (1982g) Histological study on an unusual cat muscle spindle deficient in motor innervation. Anat Embryol (Berl) 165:39–49

Kucera J, Dorovini-Zis K (1979) Types of human intrafusal muscle fibers. Muscle Nerve 2:437–451

Kucera J, Hughes R (1983) Histological study of motor innervation to long nuclear chain intrafusal fibers in the muscle spindle of the cat. Cell Tissue Res 228:535–547

Kuffler SW, Hunt CC (1952) The mammalian small-nerve fibres: a system for efferent nervous regulation of muscle spindle discharge. Res Publ Assoc Res Nerv Ment Dis 30:24–47

Kuffler SW, Vaughan-Williams EM (1953) Properties of the slow skeletal muscle fibres of the frog. J Physiol 121:318–340

Kuffler SW, Hunt CC, Quilliam JP (1951) Function of medullated small-nerve fibres in mammalian ventral roots: efferent muscle spindle innervation. J Neurophysiol 14:29–54

Landon DN (1972) The fine structure of the equatorial regions of developing muscle spindles in the rat. J Neurocytol 1:189–210

Lännergren J (1971) The effect of low-level activation on the mechanical properties of isolated frog muscle fibres. J Gen Physiol 58:145–162

Laporte Y (1979) Innervation of cat muscle spindles by fast-conducting skeleto-fusimotor fibres. In: Asanuma H, Wilson VJ (eds) Integration in the nervous system. Igaku-Shoin, Tokyo, pp 3–12

Laporte Y, Emonet-Dénand F (1973) Evidence for common innervation of bag and chain muscle fibres in cat spindles. In: Stein RB, Pearson KG, Smith RS, Redford JB (eds) Control of posture and locomotion. Plenum, New York, pp 119–126

Laporte Y, Emonet-Dénand F (1976) The skeleto-fusimotor innervation of cat muscle spindle. Prog Brain Res 44:99–106

Laporte Y, Emonet-Dénand F, Jami L (1981) The skeletofusimotor or β-innervation of mammalian muscle spindles. TINS 4:97–99

Larson CR, Smith A, Luschei ES (1981) Discharge characteristics and stretch sensitivity of jaw muscle afferents in the monkey during controlled isometric bites. J Neurophysiol 46:130–142

Larson CR, Finocchio DV, Smith A, Luschei ES (1983) Jaw muscle afferent firing during an isotonic jaw-positioning task in the monkey. J Neurophysiol 50:61–73

Laursen AM, Wiesendanger M (1966) Pyramidal effect on alpha and gamma motoneurons. Acta Physiol Scand 67:165–172

Leksell L (1945) The action potential and excitatory effects of the small ventral root fibres to skeletal muscle. Acta Physiol Scand (Suppl 31) 10:1–84

Lennerstrand G (1968a) Position and velocity sensitivity of muscle spindles in the cat. I. Primary and secondary endings deprived of fusimotor activation. Acta Physiol Scand 73:281–299

Lennerstrand G (1968b) Position and velocity sensitivity of muscle spindles in the cat. IV. Interaction between two fusimotor fibres converging on the same spindle ending. Acta Physiol Scand 74:257–273
Lennerstrand G, Thoden U (1968a) Position and velocity sensitivity of muscle spindles in the cat. II. Dynamic fusimotor single-fibre activation of primary endings. Acta Physiol Scand 74:16–29
Lennerstrand G, Thoden U (1968b) Position and velocity sensitivity of muscle spindles in the cat. III. Static fusimotor single-fibre activation of primary and secondary endings. Acta Physiol Scand 74:30–49
Lennerstrand G, Thoden U (1968c) Muscle spindle responses to concomitant variations in length and in fusimotor activation. Acta Physiol Scand 74:153–165
Lewis DM, Proske U (1972) The effect of muscle length and rate of fusimotor stimulation on the frequency of discharge in primary endings from muscle spindles in the cat. J Physiol 222:511–535
Lewis MMcD, Prochazka A, Sontag KH, Wand P (1979a) Discharge rates of muscle afferents during voluntary movements of different speeds. Prog Brain Res 50:155–162
Lewis MMcD, Prochazka A, Sontag KH, Wand P (1979b) Efferent and afferent responses during falling and landing in cat. Prog Brain Res 50:423–428
Lloyd DPC, Chang HT (1948) Afferent fibres in muscle nerves. J Neurophysiol 11: 199–208
Loeb GE, Duysens J (1979) Activity patterns of individual hindlimb primary and secondary spindle afferents during normal movements in unrestrained cats. J Neurophysiol 42:420–440
Loeb GE, Hoffer JA (1981) Muscle spindle function during normal and perturbed locomotion in cats. In: Taylor A, Prochazka A (eds) Muscle receptors and movement. MacMillan, London, pp 219–228
Lund JP, Richmond FJR, Touloumis C, Patry Y, Lamarre Y (1978) The distribution of Golgi tendon organs and muscle spindles in masseter and temporalis muscles in the cat. Neuroscience 3:259–270
Lund JP, Smith AM, Sessle BJ, Murakami T (1979) Activity of trigeminal α- and γ-motoneurons and muscle afferents during performance of a biting task. J Neurophysiol 42:710–725
Lundberg A (1979a) Integration in a propriospinal motor centre controlling the forelimb in the cat. In: Asanuma H, Wilson VJ (eds) Integration in the nervous system. Igaku-Shoin, Tokyo, pp 47–65
Lundberg A (1979b) Multisensory control of spinal reflex pathways. Prog Brain Res 50:11–28
Lundberg A, Voorhoeve PE (1962) Effects from the pyramidal tract on spinal reflex arcs. Acta Physiol Scand 56:201–219
Luschei ES, Goldberg LJ (1981) Neural mechanisms of mandibular control: mastication and voluntary biting. In: Brookhart JM, Mountcastle VB, Brooks VB (eds) Motor control, part 2. American Physiological Society, Bethesda, pp 1237–1274 (Handbook of physiology, sect 1, vol 2)
Matsunami K, Kubota K (1972) Muscle afferents of trigeminal mesencephalic tract nucleus and mastication in chronic monkeys. Jpn J Physiol 22:545–555
Matthews BHC (1931) The response of a muscle spindle during active contraction of a muscle. J Physiol 72:153–174
Matthews BHC (1933) Nerve endings in mammalian muscle. J Physiol 78:1–53
Matthews PBC (1962) The differentiation of two types of fusimotor fibre by their effects on the dynamic response of muscle spindle primary endings. Q J Exp Physiol 47:324–333
Matthews PBC (1963) The response of deefferented muscle spindle receptors to stretching at different velocities. J Physiol 168:660–678

Matthews PBC (1964) Muscle spindles and their motor control. Physiol Rev 44:219–288
Matthews PBC (1969) Evidence that the secondary as well as the primary endings of the muscle spindles may be responsible for the tonic stretch reflex of the decerebrate cat. J Physiol 204:365–393
Matthews PBC (1972) Mammalian muscle receptors and their central action. Arnold, London
Matthews PBC (1981a) Evolving views on the internal operation and functional role of the muscle spindle. J Physiol 320:1–30
Matthews PBC (1981b) Muscle spindles: their messages and their fusimotor supply. In: Brookhart JM, Mountcastle VB, Brooks VB (eds) Motor control, part 1. American Physiological Society, Bethesda, pp 189–228 (Handbook of physiology, sect 1, vol 2)
Matthews PBC (1981c) Proprioceptors and the regulation of movement. In: Towe AL, Luschei ES (eds) Motor coordination. Plenum, New York, pp 93–137 (Handbook of behavioral neurobiology, vol 5)
Matthews PBC (1982) Where does Sherrington's "muscular sense" originate? Muscles, joints, corollary discharges? Annu Rev Neurosci 5:189–218
Matthews PBC, Stein RB (1969a) The sensitivity of muscle spindle afferents to small sinusoidal changes of length. J Physiol 200:723–743
Matthews PBC, Stein RB (1969b) The regularity of primary and secondary muscle spindle afferent discharges. J Physiol 202:59–82
Matthews PBC, Watson JDG (1981) Action of vibration on the response of cat muscle spindle Ia afferents to low-frequency sinusoidal stretching. J Physiol 317:365–381
Matthews PBC, Westbury DR (1965) Some effects of fast and slow motor fibres on muscle spindles of the frog. J Physiol 178:178–192
May O, Horsley V (1910) The mesencephalic root of the fifth nerve. Brain 33:175–203
McCloskey DI (1978) Kinesthetic sensibility. Physiol Rev 58:763–820
McCloskey DI (1981) Corollary discharges: motor commands and perception. In: Brookhart JM, Mountcastle VB, Brooks VB (eds) Motor control, part 2. American Physiological Society, Bethesda, pp 1415–1447 (Handbook of physiology, sect 1, vol 2)
McIntyre AK, Proske U, Tracey DJ (1978) Fusimotor responses to volleys in joint and interosseus afferents in the cat's hindlimb. Neurosci Lett 10:287–292
McKeon B, Burke D (1981) Component of muscle spindle discharge related to arterial pulse. J Neurophysiol 46:788–796
McKeon B, Burke D (1983) Muscle spindle discharge in response to contraction of single motor units. J Neurophysiol 49:291–302
McWilliam PN (1975) The incidence and properties of beta axons to muscle spindles in the cat hindlimb. Q J Exp Physiol 60:25–36
Mense S (1977) Nervous outflow from skeletal muscle following chemical noxious stimulation. J Physiol 267:75–88
Mense S, Stahnke M (1978) Discharges in muscular group III and IV afferents during contractions of the cat's triceps surae muscle. Pflügers Arch 373:69
Mense S, Stahnke M (1983) Responses in muscle afferent fibres of slow conduction velocity to contractions and ischaemia in the cat. J Physiol 342:383–397
Merton PA (1951) The silent period in a muscle of the human hand. J Physiol 114:183–198
Merton PA (1953) Speculations on the servo-control of movement. In: Malcolm JL, Gray JAB, Wolstenholme GEW (eds) The spinal cord. Churchill, London, pp 247–255
Meyer-Lohmann J, Riebold W, Robrecht D (1974) Mechanical influence of the extrafusal muscle on the static behaviour of deafferented primary muscle spindle endings in cat. Pflügers Arch 352:267–278
Milburn A (1973) The early development of muscle spindles in the rat. J Cell Sci 12:175–195

Morimoto T, Inoue H, Kawamura Y (1982) Diameter spectra of sensory and motor fibres in nerves to jaw-closing and jaw-opening muscles in the cat. Jpn J Physiol 32: 171–182
Mortimer EM, Akert K (1961) Cortical control and representation of fusimotor neurons. Am J Phys Med 40:228–248
Murphy PR (1981) The recruitment order of γ-motoneurones in the decerebrate rabbit. J Physiol 315:59–67
Murthy KSK (1978) Vertebrate fusimotor neurones and their influences on motor behaviour. Prog Neurobiol 11:249–307
Murthy KSK (1983) Physiological identification of static β-axons in primate muscle. Exp Brain Res 52:6–8
Murthy KSK, Letbetter WD, Eidelberg E, Cameron WE, Petit J (1982) Histochemical evidence for the existence of skeletomotor (β) innervation in the primate. Exp Brain Res 46:186–190
Newsom Davis J (1975) The response to stretch of human intercostal muscle spindles studied in vitro. J Physiol 249:561–579
Nordh E, Hulliger M, Vallbo ÅB (1983) The variability of inter-spike intervals of human spindle afferents in relaxed muscles. Brain Res 271:89–99
Noth J (1971) Recurrente Hemmung der Extensor-Fusimotoneurone? Pflügers Arch 329:23–33
Noth J (1981) Autogenetic and antagonistic group II effects on extensor gamma motoneurons of the decerebrate cat. In: Taylor A, Prochazka A (eds) Muscle receptors and movement. MacMillan, London, pp 207–214
Noth J (1983) Autogenetic inhibition of extensor γ-motoneurones revealed by electrical stimulation of group I fibres. J Physiol 342:51–65
Noth J, Thilmann A (1980) Autogenetic excitation of extensor γ-motoneurones by group II muscle afferents in the cat. Neurosci Lett 17:23–26
Ovalle WK, Smith RS (1972) Histochemical identification of three types of intrafusal muscle fibres in the cat and monkey based on the myosin ATPase reaction. Can J Physiol Pharmacol 50:195–202
Paintal AS (1960) Functional analysis of group III afferent fibres of mammalian muscles. J Physiol 152:250–270
Passatore M, Filippi GM (1981) On whether there is a direct sympathetic influence on jaw muscle spindles. Brain Res 219:162–165
Passatore M, Filippi GM (1982) A dual effect of sympathetic nerve stimulation on jaw muscle spindles. J Auton Nerv Syst 6:347–361
Perret C, Berthoz A (1973) Evidence of static and dynamic fusimotor actions on the spindle response to sinusoidal stretch during locomotor activities in the cat. Exp Brain Res 18:178–188
Perret C, Buser P (1972) Static and dynamic fusimotor activity during locomotor movements in the cat. Brain Res 40:165–169
Petit J, Cameron WE, Murthy KSK, Barone LA (1983) Patterns of fusimotor innervation by γ-efferents in cat peroneus tertius. Exp Brain Res 51:146–152
Phillips CG (1969) The Ferrier Lecture, 1968. Motor apparatus of the baboon's hand. Proc R Soc Lond (Biol) 173:141–174
Poppele RE (1973) Systems approach to the study of muscle spindles. In: Stein RB, Pearson KG, Smith RS, Redford JB (eds) Control of posture and locomotion. Plenum, New York, pp 127–146
Poppele RE (1981) An analysis of muscle spindle behavior using randomly applied stretches. Neuroscience 6:1157–1165
Poppele RE, Bowman RJ (1970) Quantitative description of linear behavior of mammalian muscle spindle. J Neurophysiol 33:59–72
Poppele RE, Kennedy WR (1974) Comparison between behaviour of human and cat muscle spindles recorded in vitro. Brain Res 75:316–319
Poppele RE, Quick DC (1981) Stretch-induced contraction of intrafusal muscle in cat muscle spindles. J Neurosci 1:1069–1074

Poppele RE, Kennedy WR, Quick DC (1979) A determination of static mechanical properties of intrafusal muscle in isolated cat muscle spindles. Neuroscience 4: 401–411

Post EM, Rymer WZ, Hasan Z (1980) Relation between intrafusal and extrafusal activity in triceps surae muscles of the decerebrate cat: evidence for β-action. J Neurophysiol 44:383–404

Pringle JWS (1949) The excitation and contraction of the flight muscles of insects. J Physiol 108:226–232

Pringle JWS (1978) Stretch activation of muscle: function and mechanisms. Proc R Soc Lond (Biol) 201:107–130

Prochazka A (1981) Muscle spindle function during normal movement. Int Rev Physiol 5:47–90

Prochazka A, Hulliger M (1983) Muscle afferent function and its significance for motor control mechanisms during voluntary movements in cat, monkey and man. In: Desmedt JE (ed) Motor control mechanisms in health and disease. Raven, New York, pp 93–132

Prochazka A, Wand P (1981) Independence of fusimotor and skeletomotor systems during voluntary movement. In: Taylor A, Prochazka A (eds) Muscle receptors and movement. MacMillan, London, pp 229–244

Prochazka A, Westerman RA, Ziccone SP (1976) Discharges of single hindlimb afferents in freely moving cat. J Neurophysiol 39:1090–1104

Prochazka A, Westerman RA, Ziccone SP (1977) Ia afferent activity during a variety of voluntary movements in the cat. J Physiol 268:423–448

Prochazka A, Stephens JA, Wand P (1979) Muscle spindle discharge in normal and obstructed movements. J Physiol 287:57–66

Proske U (1975) Stretch-evoked potentiation of responses of muscle spindles in the cat. Brain Res 88:378–383

Proske U, Lewis DM (1972) The effects of muscle stretch and vibration on fusimotor activity in the lightly anaesthetized cat. Brain Res 46:55–69

Quick DC, Kennedy WR, Donaldson L (1979) Dimensions of myelinated nerve fibres near the motor and sensory terminals in cat tenuissimus muscles. Neuroscience 4: 1089–1096

Quick DC, Kennedy WR, Poppele RE (1980) Anatomical evidence of multiple sources of action potentials in the afferent fibres of muscle spindles. Neuroscience 5:109–115

Rack PMH, Westbury DR (1966) The effects of suxamethonium and acetylcholine on the behaviour of cat muscle spindles during dynamic stretching, and during fusimotor stimulation. J Physiol 186:698–713

Rakhawy MT, Shehata SH, Badawy ZH (1971) The proprioceptive innervation of the lateral pterygoid muscle in man and some other mammals. Acta Anat (Basel) 79: 581–598

Renshaw B (1941) Influence of discharge of motoneurons upon excitation of neighbouring motoneurons. J Neurophysiol 4:167–183

Renshaw B (1946) Central effects of centripetal impulses in axons of spinal ventral roots. J Neurophysiol 9:191–204

Rexed B, Therman PO (1948) Calibre spectra of motor and sensory nerve fibres to flexor and extensor muscles. J Neurophysiol 11:133–139

Richmond FJR, Abrahams VC (1975) Morphology and distribution of muscle spindles in dorsal muscles of the cat neck. J Neurophysiol 38:1322–1339

Richmond FJR, Abrahams VC (1979a) Physiological properties of muscle spindles in dorsal neck muscles of the cat. J Neurophysiol 42:604–617

Richmond FJR, Abrahams VC (1979b) What are the proprioceptors of the neck? Progr Brain Res 50:245–254

Richmond FJR, Bakker DA (1982) Anatomical organization and sensory receptor content of soft tissues surrounding upper cervical vertebrae in the cat. J Neurophysiol 48:49–61

Richmond FJR, Anstee GCB, Sherwin EA, Abrahams VC (1976) Motor and sensory fibres of neck muscle nerves in the cat. Can J Physiol Pharmacol 54:294–304
Roll JP, Vedel JP (1982) Kinaesthetic role of muscle afferents in man, studied by tendon vibration and microneurography. Exp Brain Res 47:177–190
Rosenthal NP, McKean TA, Roberts WJ, Terzuolo CA (1970) Frequency analysis of stretch reflex and its main subsystems in triceps surae muscles of the cat. J Neurophysiol 33:713–749
Rossi G (1927) Asimmetrie toniche posturali, ed asimmetrie motorie. Arch Fisiol 25: 146–157
Ruffini A (1897) Observations on sensory nerve endings in voluntary muscles. Brain 20:368–374
Ruffini A (1898) On the minute anatomy of the neuromuscular spindles of the cat, and on their physiological significance. J Physiol 23:190–208
Saito M, Tomonaga M, Hirayama K, Narabayashi H (1977) Histochemical study of normal human muscle spindle. Histochemical classification of intrafusal muscle fibres and intrafusal nerve endings. J Neurol 216:79–89
Sato H (1981) Fusimotor modulation by spinal and skin temperature changes and its significance in cold shivering. Exp Neurol 74:21–32
Sato H (1983) Effects of skin cooling and warming on stretch responses of the muscle spindle primary and secondary afferent fibres from the cat's tibialis anterior. Exp Neurol 81:446–458
Sato H, Hasegawa Y (1977) Reflex changes in discharge activities of gamma efferents to varying skin temperatures in cats. Pflügers Arch 372:195–201
Schäfer SS (1973) The characteristic curves of the dynamic response of primary muscle spindle endings in the absence and presence of stimulation of fusimotor fibres. Brain Res 59:395–399
Schäfer SS (1974) The discharge frequencies of primary muscle spindle endings during simultaneous stimulation of two fusimotor filaments. Pflügers Arch 350:359–372
Schäfer SS, Schäfer S (1973) The behaviour of the proprioceptors of the muscle and the innervation of the fusimotor system during cold shivering. Exp Brain Res 17: 364–380
Schieber MH, Thach WT (1980) Alpha-gamma dissociation during slow tracking movements of the monkey's wrist: preliminary evidence from spinal ganglion recording. Brain Res 202:213–216
Schmidt RF (1978) Fundamentals of neurophysiology, 2nd edn. Springer, Berlin Heidelberg New York
Schmidt RF, Kniffki KD, Schomburg ED (1981) Der Einfluß kleinkalibriger Muskelafferenzen auf den Muskeltonus. In: Bauer HJ, Koella WP, Struppler A (eds) Therapie der Spastik. Verlag für angewandte Wissenschaften, München, pp 71–86
Sears TA (1963) Activity of fusimotor fibres innervating muscle spindles in the intercostal muscles of the cat. Nature 197:1013–1014
Sears TA (1964) Efferent discharges in alpha- and fusimotor fibres of intercostal nerves of the cat. J Physiol 174:295–315
Sears TA (1973) Servo control of the intercostal muscles. In: Desmedt JE (ed) New developments in electromyography and clinical neurophysiology, vol 3. Karger, Basel, pp 404–417
Severin FV (1970) The role of the gamma motor system in the activation of the extensor alpha motoneurones during controlled locomotion. Biophysics 15:1138–1145
Severin FV, Orlovsky GN, Shik ML (1967) Work of the muscle spindle receptors during controlled locomotion. Biophysics 12:575–586
Sherrington CS (1894) On the anatomical constitution of nerves of skeletal muscles: with remarks on recurrent fibres in the ventral spinal nerve-root. J Physiol 17:211–258
Shimazu H, Hongo T, Kubota K (1962a) Two types of central influences on gamma motor system. J Neurophysiol 25:309–323

Shimazu H, Hongo T, Kubota K (1962b) Nature of central regulation of muscle spindle activity. In: Barker D (ed) Symposium on muscle receptors. Hong Kong University Press, Hong Kong, pp 49–57

Sjöström A, Zangger P (1975) α-γ Linkage in the spinal generator for locomotion of the cat. Acta Physiol Scand 94:130–132

Sjöström A, Zangger P (1976) Muscle spindle control during locomotor movements generated by the deafferented spinal cord. Acta Physiol Scand 97:281–291

Smith JL, Hutten RS, Eldred E (1974) Postcontraction changes in sensitivity of muscle afferents to static and dynamic stretch. Brain Res 78:193–202

Smith RD, Marcarian HQ, Niemer WT (1968) Direct projections from the masseteric nerve to the mesencephalic nucleus. J Comp Neurol 133:495–502

Smith RS, Ovalle WK (1972) The structure and function of intrafusal muscle fibres. Muscle Biol 1:147–227

Stauffer EK, Watt DGD, Taylor A, Reinking RM, Stuart DG (1976) Analysis of muscle receptor connections by spike-triggered averaging. 2. Spindle group II afferents. J Neurophysiol 39:1393–1402

Stein RB (1974) Peripheral control of movement. Physiol Rev 54:215–243

Struppler A, Velho F (1976) Single muscle spindle afferent recordings in human flexor reflex. In: Shahani M (ed) Motor systems: neurophysiology and muscle mechanisms, vol 4. Elsevier, Amsterdam, pp 197–207

Swash M, Fox KP (1972) Muscle spindle innervation in man. J Anat 112:61–80

Szentagothai J (1948) Anatomical considerations of monosynaptic reflex arcs. J Neurophysiol 11:445–454

Szumski AJ, Burg D, Struppler A, Velho F (1974) Activity of muscle spindles during muscle twitch and clonus in normal and spastic human subjects. Electroencephalogr Clin Neurophysiol 37:589–597

Taylor A (1981) A critique of the papers by Vallbo and Hagbarth. In: Taylor A, Prochazka A (eds) Muscle receptors and movement. MacMillan, London, pp 287–290

Taylor A, Appenteng K (1981) Distinctive modes of static and dynamic fusimotor drive in jaw muscles. In: Taylor A, Prochazka A (eds) Muscle receptors and movement. MacMillan, London, pp 179–192

Taylor A, Cody FWJ (1974) Jaw muscle spindle activity in the cat during normal movements of eating and drinking. Brain Res 71:523–530

Taylor A, Davey MR (1968) Behaviour of jaw muscle stretch receptors during active and passive movements in the cat. Nature 220:301–302

Taylor A, Prochazka A (1981) Muscle receptors and movement. MacMillan, London

Te Kronnie G, Donselaar Y, Soukup T, Zelená J (1982) Development of immunohistochemical characteristics of intrafusal fibres in normal and de-efferented rat muscle spindles. Histochemistry 74:355–366

Thelander HE (1924) The course and distribution of the radix mesencephalica trigemini in the cat. J Comp Neurol 37:207–220

Thorson J, Biederman-Thorson M (1974) Distributed relaxation processes in sensory adaptation. Science 183:161–172

Trott J (1976) The effect of low-amplitude muscle vibration on the discharge of fusimotor neurones in the decerebrate cat. J Physiol 255:635–649

Vallbo ÅB (1970a) Slowly adapting muscle receptors in man. Acta Physiol Scand 78: 315–333

Vallbo ÅB (1970b) Discharge patterns in human muscle spindle afferents during isometric voluntary contractions. Acta Physiol Scand 80:552–566

Vallbo ÅB (1971) Muscle spindle response at the onset of voluntary isometric contractions in man. Time difference between fusimotor and skeletomotor effects. J Physiol 218:405–431

Vallbo ÅB (1972) Single unit recording from human peripheral nerves: muscle receptor discharge in resting muscles and during voluntary contractions. Excerpta Med Int Congr Ser 253:283–297

Vallbo ÅB (1973a) The role of the muscle spindles in voluntary motor activity. Acta Physiol Scand (Suppl) 396:38

Vallbo ÅB (1973b) Muscle spindle afferent discharge from resting and contracting muscles in normal human subjects. In: Desmedt JE (ed) New developments in electromyography and clinical neurophysiology, vol 3. Karger, Basel, pp 251–262

Vallbo ÅB (1974a) Afferent discharge from human muscle spindles in non-contracting muscles. Steady state impulse frequency as a function of joint angle. Acta Physiol Scand 90:303–318

Vallbo ÅB (1974b) Human muscle spindle discharge during isometric voluntary contractions. Amplitude relations between spindle frequency and torque. Acta Physiol Scand 90:319–336

Vallbo ÅB (1981) Basic patterns of muscle spindle discharge in man. In: Taylor A, Prochazka A (eds) Muscle receptors and movement. MacMillan, London, pp 263–276

Vallbo ÅB, Hagbarth K-E (1968) Activity from skin mechanoreceptors recorded percutaneously in awake human subjects. Exp Neurol 21:270–289

Vallbo ÅB, Hagbarth K-E (1973) Micro-electrode recordings from human peripheral nerves. In: Desmedt JE (ed) New developments in electromyography and clinical neurophysiology, vol 2. Karger, Basel, pp 67–84

Vallbo ÅB, Hulliger M (1979) Flexible balance between skeletomotor and fusimotor activity during voluntary movements in man. Neurosci Lett (Suppl) 3:103

Vallbo ÅB, Hulliger M (1981) Independence of skeletomotor and fusimotor activity in man? Brain Res 223:176–180

Vallbo ÅB, Hulliger M (1982) The dependence of discharge rate of spindle afferent units on the size of the load during isotonic position holding in man. Brain Res 237:297–307

Vallbo ÅB, Hagbarth K-E, Torebjörk HE, Wallin BG (1979) Somatosensory, proprioceptive, and sympathetic activity in human peripheral nerves. Physiol Rev 59: 919–957

Vedel JP (1975) Effets des stimulations du noyau caudé sur l'activité des terminaisons fusoriales primaires et secondaires du muscle soléaire. Exp Brain Res 22:113–128

Vedel JP, Coulmance M (1975) Modulation de l'activité des terminaisons fusoriales primaires et secondaires du muscle tibial antérieur au cours de contractions induites par stimulations du novau caudé. Exp Brain Res 22:129–144

Vedel JP, Mouillac-Baudevin J (1969a) Contrôle de l'activité des fibres fusimotrices dynamiques et statiques par la formation réticulée mésencéphalique chez le chat. Exp Brain Res 9:307–324

Vedel JP, Mouillac-Baudevin J (1969b) Étude fonctionelle du contrôle de l'activité des fibres fusimotrices dynamiques et statiques par les formations réticulées mésencéphaliques, pontiques et bulbaires chez la chat. Exp Brain Res 9:325–345

Vedel JP, Mouillac-Baudevin J (1970) Contrôle pyramidal de l'activité des fibres fusimotrices dynamiques et statiques chez le chat. Exp Brain Res 10:39–63

Vedel JP, Paillard J (1965) Effet différentiel des stimulations du noyau caudé et du cortex frontal sur la sensibilité dynamique des terminaisons fusoriales primaires chez le chat. J Physiol (Paris) 57:716–717

Voorhoeve PE, Van Kanten RW (1962) Reflex behaviour of fusimotor neurones of the cat upon electrical stimulation of various afferent fibres. Acta Physiol Pharmacol Neerl 10:391–407

Wallin BG, Hagbarth K-E (1978) Muscle spindle activity in man during voluntary alternating movements, parkinsonian tremor and clonus. In: Desmedt JE (ed) Progress in clinical neurophysiology, vol 5. Karger, Basel, pp 150–159

Wand P, Schwarz M, Kolasiewicz W, Sontag KH (1981) Nigral output neurons are engaged in regulation of static fusimotor action onto flexors in cat. Pflügers Arch 391:255–257

Westbury D (1964) A quantitative study of muscle proprioceptors. B Sc Thesis, Oxford University

Westbury DR (1980) Lack of a contribution from gamma motoneurone axons to Renshaw inhibition in the cat spinal cord. Brain Res 186:217–221

Westbury DR (1982) A comparison of the structures of α and γ spinal motoneurones of the cat. J Physiol 325:79–91

Wilson VJ (1966) Regulation and function of Renshaw cell discharge. In: Granit R (ed) Muscular afferents and motor control, Nobel symposium. Almqvist and Wiksell, Stockholm, pp 317–329

Windhorst U, Meyer-Lohmann J (1977) The influence of extrafusal muscle activity on the discharge patterns of primary spindle endings. Pflügers Arch 372:131–138

Windhorst U, Meyer-Lohmann J, Schmidt J (1975) Correlation of the dynamic behaviour of de-efferented primary muscle spindle endings with their static behaviour. Pflügers Arch 357:113–122

Windhorst U, Schmidt J, Meyer-Lohmann J (1976) Analysis of the dynamic responses of de-efferented primary muscle spindle endings to ramp stretch. Pflügers Arch 366: 233–240

Wuerker RB, Henneman E (1963) Reflex regulation of primary (annulospiral) stretch receptors via gamma motoneurons in the cat. J Neurophysiol 26:539–550

Yanagisawa N, Narabayashi H, Shimazu H (1963) Thalamic influences on the gamma motor system. Arch Neurol 9:348–357

Yokota T, Voorhoeve PE (1969) Pyramidal control of fusimotor neurones supplying extensor muscles in the cat's forelimb. Exp Brain Res 9:96–115

Young RR, Hagbarth K-E (1980) Physiological tremor enhanced by manoeuvres affecting the segmental stretch reflex. J Neurol Neurosurg Psychiatry 43:248–256

Zelená J (1964) Development, degeneration and regeneration of receptor organs. Progr Brain Res 13:175–213

Zelená J, Hnik P (1963) Effect of innervation on the development of muscle receptors. In: Gutmann E, Hnik P (eds) The effect of use and disuse on neuromuscular functions. Elsevier, Amsterdam, pp 95–105

Zelená J, Soukup T (1973) Development of muscle spindles deprived of fusimotor innervation. Z Zellforsch 144:435–452

Zelená J, Soukup T (1974) The differentiation of intrafusal fibre types in rat muscle spindles after motor denervation. Cell Tissue Res 153:115–136

Rev. Physiol. Biochem. Pharmacol., Vol. 101

Interaction of Viruses with Tumor Promotors

NAOKI YAMAMOTO

Contents

1 Introduction

Carcinogenesis is usually a multistep process and it is likely that most human cancers result from a complex interaction between both environmental and endogenous factors (*Hiatt* et al. 1977; *Weinstein* et al. 1979).

Institute for Virus Research, Kyoto University, Sakyo-ku, Kyoto, Japan 606 (Present address: Dept. of Virology and Parasitology, Yamaguchi Univ. School of Med., Ube, 755 Japan)

The best-defined experimental system for multistage carcinogenesis is the mouse skin, in which one can clearly identify two distinct stages, one termed "initiation" and the other "promotion" (*Van Duuren* and *Orris* 1965; *Boutwell* 1974; *Hecker* 1975; *Berenblum* 1975; *Slaga* et al. 1978). The agents responsible for each stage are termed tumor initiators and tumor promoters, respectively. Several pioneering studies had been done previously, but Berenblum was the first to establish the modern concept of two-stage or multistep carcinogenesis (*Berenblum* 1954). From the results of his experiments in painting mouse skin, it was possible to calculate the relationships between both the initiation step and the promotion step and between initiators and promoters. Initiators are carcinogenic on their own, when added in high concentrations or used repeatedly in smaller concentrations. An initiator such as benzo(a)pyrene does not cause skin tumors if a single subcarcinogenic dose is painted on the skin. Skin tumors, however, are produced when croton oil, a tumor promoter extracted from a plant, *Croton tiglium,* is used after the application of an initiator. Later, the active component in croton oil was determined by *Hecker* et al. (1968) and *Van Duuren* (1969) to be a phorbol diester, 12-0-tetradecanoylphorbol-13-acetate TPA. Several other types of tumor promoters have been found subsequently (reviewed in *Hecker* 1978).

A tumor promoter is neither mutagenic nor carcinogenic on its own, but it exerts the effect when used after the application of initiators. If the order of application of the two chemicals is reversed, they do not cause cancer. Most initiating agents first require metabolic actions in the cells in order to become active, while tumor promoters do not require such actions. Electrophiles yielded by the activation of initiators bind covalently to macromolecules such as DNA, RNA, and proteins in the

---

The following abbreviations appear in the text:

acyclovir (9-(2-hydroxyethoxymethyl)-guanine), Ad5 (adenovirus type 5), AMD (actinomycin D), AO (acridine orange), ATL (adult T-cell leukemia), ATLV (ATL virus), BL (Burkitt's lymphoma), BPV-1 (bovine papilloma virus type 1), CBL (cord blood lymphocytes), cDNA (complementary DNA), CH (cycloheximide), C-kinase (Ca^{2+}-activated, phospholipid-dependent protein kinase), EA (early antigens), EBNA (EBV-associated nuclear antigens), EBV (Epstein-Barr virus), EGF (epidermal growth factor), ETB (ethidium bromide), FLV (Friend leukemia virus), HCS (human cord serum), HEK (human embryonic kidney), IM (infectious mononucleosis), IUdR (iododeoxyuridine), LCL (lymphoblastoid cell line), MG (methyl green), MLV (murine leukemia virus), MMC (mitomycin C), MNNG (N-methyl-N'nitro-nitrosoguanidine), MPMV (Mason-Pfizer monkey virus), MSV (murine sarcoma virus), NPC (nasopharyngeal carcinoma), NRK (normal rat kidney), ODC (ornithine decarboxylase), PA (plasminogen activator), PBS (phosphate-buffered saline), PDD (phorbol-12,13-didecanoate), RSV (Rous sarcoma virus), SSV (simian sarcoma virus), TPA (12-0-tetradecanoyl-phorbol-13-acetate), ts (temperature-sensitive), TX (trioxsalen), VCA (viral capsid antigens) cells. No evidence has been presented for covalent binding of promoting agents to the macromolecules.

One of the most striking characteristics of promoting agents is their reversibility. In the mouse two-stage skin carcinogenesis model, tumors first appear as papillomas, and repeated painting with promoters results in the appearance of carcinomas. Benign tumors (papillomas) will regress if the promoter is discontinued at this stage. In contrast to promoting agents, initiating agents are irreversible. Even if the initiating agents have been discontinued for a long time, the application of promoters causes hyperplasia in mouse skin. This suggests that after treatment with the initiators, the affected cells remain dormant in the skin.

Epstein-Barr virus (EBV) has been associated with two human malignancies, African Burkitt's lymphoma (BL) and nasopharyngeal carcinoma (NPC) (reviewed in *Klein* 1973; *zur Hausen* 1975; *Epstein* and *Achong* 1977). It also causes infectious mononucleosis (IM) (*Henle* et al. 1968; *Henle* and *Henle* 1974). The presence of EBV-specific DNA in cells of both types of malignant tumor has been often demonstrated by nucleic acid hybridization (*zur Hausen* and *Schulte-Holthausen* 1970; *zur Hausen* et al. 1970; *Nonoyama* and *Pagano* 1971; *Nonoyama* et al. 1973; *Wolf* et al. 1973). Apparently, all cells carrying the EBV genome express the EB viral nuclear antigen (EBNA), detectable by anticomplement immunofluorescence (*Reedman* and *Klein* 1973). It has been established that certain strains of EBV readily transform human B lymphocytes in vitro into lymphoblastoid cell lines (LCL) with unlimited growth (*Henle* et al. 1967; *Pope* et al. 1968; *Yamamoto* et al. 1976).

Continuous lymphoblastoid cell lines derived from BL or from patients with IM vary widely in the expression of viral antigens and in the production of infectious viral particles, although all of them harbor the genome. In the majority of lines, spontaneous induction of viral antigens and particle synthesis occur at a low rate. Cells of the nonproducer lines, such as the Raji cell line of BL origin (*Pulvertaft* 1965), express only the nuclear antigen EBNA, but do not synthesize at detectable levels antigens associated with the replication of EBV, such as the early antigen (EA) (*Henle* et al. 1970) or the viral capsid antigen (VCA) (*zur Hausen* et al. 1967). Producer lines, however, such as the P3HR-1 line of BL origin (*Hinuma* et al. 1967), reveal a relatively constant proportion of cells containing viral capsid antigens and releasing particles.

EBV-specific DNA is present in multiple genome equivalents per cell in all EBV-transformed lymphoblastoid lines (*zur Hausen* and *Schulte-Holthausen* 1970; *Nonoyama* and *Pagano* 1971; *zur Hausen* et al. 1972; *Pritchett* et al. 1976). Most of the viral DNA persists in an episomal state as covalently closed circular molecules (*Nonoyama* and *Pagano* 1972b; *Lin-*

dahl et al. 1976; *Kashka-Dierich* et al. 1976). A smaller fraction may associate with cellular DNA in an integrated form (*Adams* et al. 1973).

EBV-inducing activity of a potent tumor promoter, TPA, and related compounds was discovered by *zur Hausen* et al. (1978) in the course of experiments initiated to study polyploidization of human lymphoblastoid cell lines by tumor promoters. Studies on the molecular biology of EBV and virus-host cell interactions have been greatly hampered in the past by the lack of a suitable tissue culture system for virus propagation and the limited amount of virus-specific components obtained from cells with a high spontaneous induction rate (*Hinuma* et al. 1967). Although the synthesis of virus-specific proteins can be enhanced substantially by the addition of bromo-and iododeoxyuridine (IUdR) (*Gerber* 1972; *Hampar* et al. 1972; *Sugawara* et al. 1972) or *n*-butyric acid (*Luka* et al. 1979), the inhibition of DNA synthesis caused by these drugs has not permitted recovery of increased yields of viral particles following purification procedures. Thus, an induction system with complete maturation of EBV particles would clearly benefit EBV research, particularly in view of the possible involvement of this virus in two human malignant tumors, BL and NPC (*Tooze* 1981).

In this review, I intend to summarize studies on the interaction between the persisting oncogenic herpes virus EBV and tumor promoters, with special reference to virus induction and the transformation of normal human lymphocytes by EBV. The effect of tumor promoters on other viral systems will also be discussed. For the interaction of tumor promoters and other viral systems, especially the adenovirus, readers are referred to the excellent reviews by *Fisher* (1982, 1983).

2 Effect of Tumor Promoters on Cells in Culture

Cell culture systems provide useful tools for the study of the complex mechanism of carcinogenesis and tumor promotion. Chemical carcinogens have been shown to interact synergistically with various viruses, such as adenovirus (*Casto* 1973; *Fisher* et al. 1978), simian virus 40 (SV40) (*Hirai* et al. 1974; *Milo* et al. 1978), and mouse leukemia virus (MLV) (*Freeman* et al. 1970; *Rhim* et al. 1971) in the transformation of cells in vitro. In 1976, *Mondal* et al. reported that in vitro transformation proceeds according to the two-stage carcinogenesis concept. They observed that phorbol esters, saccharin, or anthralin have promoting activities in C3H 10T1/2 cells derived from mouse embryo fibroblast. Thereafter, TPA was reported to enhance the transformation of cells previously exposed to a variety of initiating agents, such as chemical carcinogens

(*Lasne* et al. 1974; *Mondal* et al. 1976), UV light (*Mondal* and *Heidelberger* 1976), X-rays (*Kennedy* et al. 1978), or various oncogenic viruses, including human adenovirus 5 (Ad5) (*Fisher* et al. 1978), EBV (*Yamamoto* and *zur Hausen* 1979), or murine sarcoma virus (MSV) (*Miyaki* et al. 1980).

In addition to interaction with viral genomes, a wide variety of effects of tumor promoting agents, especially of TPA, in tissue culture cells have been reported. These effects can be divided into three categories: 1) Mimicry of transformation in normal cells and augmentation of transformed phenotype. 2) Membrane and receptor effects, including reversible effects on the morphology, refractility, and size of the monolayer cells (*Sivak* et al. 1969; *Diamond* et al. 1974b); increase of cell to cell or cell to substratum attachment (*Yamasaki* et al. 1979); inhibition of intercellular communication (*Yotti* et al. 1979); stimulation of platelet aggregation (*Zucker* et al. 1974); stimulation of incorporation of ^{3}H-choline into phospholipid in the cells (*Suss* et al. 1971); uptake of glucose, ^{32}P, and ^{86}Rb (*Driedger* and *Blumberg* 1977; *Moroney* et al. 1978); stimulation of prostaglandin synthesis, release of arachidonic acid from the cells (*Ohuchi* and *Levine* 1978); inhibition of epidermal growth factor (EGF) binding to the cells, alteration in membrane receptors (*Brown* et al. 1979; *Shoyab* et al. 1979); increase in the enzyme ornithine decarboxylase (ODC) (*Yuspa* et al. 1976); decrease in plasminogen activator (PA) (*Wigler* and *Weinstein* 1976); stimulation of DNA synthesis (*Colburn* et al. 1978; *O'Brien* et al. 1979). 3) Modulation of differentiation (inhibition or stimulation). The inhibitory effect of TPA on cell differentiation has been observed with various systems, such as murine erythroleukemia cells (*Rovera* et al. 1977; *Yamasaki* et al. 1977), mouse neuroblastoma (*Ishii* et al. 1978), chick chondroblast (*Pacifici* and *Holtzer* 1977), and 3T3 fibroblast (*Diamond* et al. 1978). Reciprocal effects of the same agent on differentiation have also been reported in some systems (*Miao* et al. 1978).

A number of studies on the effects of TPA in vitro, as mentioned above, have suggested that the cell surface membrane may be the first target of TPA action. The lipophilic nature of TPA has hindered studies on whether or not the cells contain receptors for tumor promoters (*Lee* and *Weinstein* 1978c; *Yamamoto* and *Bauer* 1981). *Driedger* and *Blumberg* (1980) first showed direct evidence of high-affinity saturable surface receptors employing ^{3}H-phorbol dibutyrate, which is much less lipophilic than TPA. Receptors for phorbol esters have been found in various types of cells from many species of animals (*Shoyab* and *Todaro* 1980). More recently, *Castagna* et al. (1982) showed that TPA directly activated in vitro Ca^{2+}-activated, phospholipid-dependent protein kinase (C-kinase) in the human platelet system without associated phosphatidyl inositol breakdown, which results in the generation of diacylglycerol, itself an activator of

C-kinase. This and subsequent studies led the authors to reason that perhaps the kinase itself was the receptor (*Niedel* et al. 1983; *Ashendel* et al. 1983). As is well known, certain polypeptide hormones and mitogens, such as EGF, platelet-derived growth factor, and insulin, have their primary effects at the cell surface and secondarily affect the nuclear and cytoplasmic functions by transmembrane signals, as mediated by the phosphorylation of serine and threonine residues in specific cellular proteins. Certain oncogenes code for protein kinases that phosphorylate tyrosine residues (*Bishop* and *Varmus* 1982). As these growth hormones and oncogenes are also able to activate C-kinase of the target cells through the breakdown of phospholipid, the interaction of these kinases and the possible presence of cascade mechanisms become important. Since TPA mimics the activities of such hormones and tumor viruses, the hypothesis that tumor promoters exert their effects via C-kinase is in agreement with the findings on growth factors, chemical carcinogens, and tumor viruses.

Under physiological conditions, diacylglycerol is produced in tissues by a reaction to exogenous signals with an extremely fast turnover rate (*Takai* 1983). In contrast, C-kinase activity enhanced by treatment with phorbol diesters is maintained at a continuously high level because this chemical is not as readily metabolized as the other compounds. This situation may provide continuous activation of cells and thus facilitate tumor development.

More extensive surveys of the luterature on the effect of tumor promoters on tissue culture cells have been made (see reviews by *Diamond* et al. 1978, 1980; *Weinstein* et al. 1978).

3 Induction of Virus Production

3.1 Induction of EBV-Specific Antigens and Viral DNA by TPA

EBV induction by TPA was initially studied with an immunofluorescence technique by *zur Hausen* et al. (1978). EBV producer P3HR-1, B95-8 (*Miller* et al. 1974), and QIMR-WIL (*Pope* et al. 1968) cell lines spontaneously express EBV-associated antigens at 3%–5%. Treatment of these three EBV producer lines and the nonproducer lines Raji and NC-37 with 20 ng/ml TPA resulted in a dramatic increase in antigen-positive cells. Both EA and VCA were increased in P3HR-1, QIMR-WIL, and B95-8, and EA only was induced in NC-37 and Raji cells. Of the producer cells, more than 50% of the P3HR-1 and B95-8 cells revealed characteristic immunofluorescence after indirect staining with EBV VCA-reactive sera, as did about 10% of the QIMR-WIL cells after 5 days of treatment. No fluorescence was observed with CVA-negative control sera. After 8–10 days, most of the P3HR-1 cells were dead, presumably due to virus production, but some

B95-8 cells survived and continued to grow. Parallel with the increase in VCA-positive cells, effective virus production could be visualized by means of electron microscopy. In NC-37 and Raji cells, about 20% and 5%, respectively, of the cells were EA positive with TPA treatment. The addition of TPA thus revealed an important discrepancy between EBV producing and nonproducing cell lines: In the former, TPA induces complete virus particle synthesis, while early function of EBV is induced exclusively in the latter.

With the dramatic increase in viral antigen synthesis, TPA has also been shown to increase viral DNA synthesis in EBV producer cell lines P3HR-1, B95-8, and QIMR-WIL (*zur Hausen* et al. 1979; *Lin* et al. 1979; *Hudewentz* et al. 1980). Patterns of increase in viral DNA synthesis were very similar to those of VCA synthesis, showing the initial increase after 48 h, with maximal induction 5 or 6 days after treatment. However, if two assays were run in parallel, viral antigen induction appeared to precede the marked increase in viral DNA. The amount of viral DNA in the producer cell lines generally increased 10- to 20-fold after TPA treatment. In the nonproducer cell lines Raji and NC-37, the amount of viral DNA per cell was constant, despite efficient EA induction by TPA (*Bister* et al. 1979; *Lin* et al. 1979; *Hudewentz* et al. 1980). This shows that in nonproducer cells, the mechanism of induction of the persisting EBV genomes does not involve specific amplification of viral DNA, although the absolute genome copy number per cell is apparently increased along with polyploidization of the cells observed after treatment with tumor promoters (*zur Hausen* et al. 1978).

Restriction endonuclease cleavage of EBV DNA obtained from B95-8 or P3HR-1 cells with BamHl, EcoRl, and Xhol revealed no significant differences in the patterns of DNA resulting from TPA-treated or nontreated cultures (*zur Hausen* et al. 1979; *Lin* et al. 1979). These data show that viral progeny from cells treated with diterpene esters apparently do not differ from viruses isolated from noninduced cultures.

An increase in EBV genome copies per cell parallels the increase in infectious virus production. EBV DNA synthesized in Raji cells superinfected by virus prepared from TPA-induced P3HR-1 cells is increased about 15 times more than that of Raji cells superinfected with control virus (*Lin* et al. 1979). Additionally, viral DNA induced by TPA was shown to be biologically active if injected into different types of cells (*Graessmann* et al., unpublished data). It should, however, be noted that the relative amount of DNA recovered from TPA-treated cultures appears to exceed the increase observed in infectivity assays. This may point to a particularly enhanced production of nonenveloped (noninfectious) particles, which agrees with electron microscopic observations (*zur Hausen,* unpublished data).

In order to further assess TPA induction, an additional 27 lymphoblastoid cell lines harboring EBV were trested for viral antigen induction 5 days after treatment by immunofluorescence. In every EBV-transformed cell line, some increase in VCA or EA synthesis was noted, although the frequency of viral antigen induction varied considerably (*zur Hausen* et al. 1982). Altogether 24 of 27 cell lines responded to the inducing drug, and three nonproducer lines were refractory to the inducer.

The nonproducer cell lines Raji and NC-37 contain an average of 50 EBV genome equivalents per cell (*Nonoyama* and *Pagano* 1973). Other chemical inducers in addition to TPA induced EA only, without any evidence for viral structural antigen synthesis and viral DNA replication. These results suggest that nonproducer cell lines are defective in endogenous viral DNA as determined by the level of polypeptide (*Kallin* and *Klein* 1983) and nucleic acid (*Skare* and *Farley* 1981) synthesis.

3.2 Induction of EBV-Specific DNA Polymerase and DNAse by TPA

Herpes simplex virus-specific DNA polymerase was first found by *Keir* and *Gold* (1963). *Weissbach* et al. (1973) compared this enzyme biochemically with the DNA polymerase in mammalian cells. The viral polymerase showed an exceptional dependency on a high salt concentration [e.g., 100–150 m*M* K_2SO_4 or $(NH_4)_2SO_4$], compared with cellular DNA polymerase. It has been demonstrated that a similar DNA polymerase was detected in P3HR-1 cells and D98/HR-1 cells after IUdR treatment (*Ooka* et al. 1979; *Miller* and *Rapp* 1977) or in superinfected Raji cells (*Clough* and *McMahon* 1981; *Feighny* et al. 1981). More recently, various groups have demonstrated a rapid induction of EBV-associated DNA polymerase activity in P3HR-1 cells after treatment with TPA (*Datta* et al. 1980b; *Clough* 1981). TPA induction of DNA polymerase was seen not only in producer P3HR-1 cells, but also in nonproducer Raji cells. This indicates that the expression of EBV-specific DNA polymerase does not require viral DNA synthesis. Induction of this enzymatic activity is specifically inhibited by various chemicals, such as phosphonoacetic acid, phosphonoformate, and arabinosylnucleosides (Ara C, Ara A, and Ara T) (*Grossberger* and *Clough* 1981; *Ooka* 1983). Furthermore, as described in a later part of this review, treatment of the cells with these chemicals did not interfere with EA induction by TPA (*Ooka* 1983). Although it is yet to be determined whether this enzyme is encoded by the EBV genome, the following data suggest that this is very likely: The enzyme 1) has a striking similarity to other herpes virus polymerases; 2) it is induced first by the activation of EBV genomes together with EA; and 3) it is not detectable in EBV genome-negative lymphoblastoid cells.

EBV-specific DNAse was also found in the virus producers B95-8 and P3HR-1 cells, superinfected Raji cells and IUdR-induced hybrid cells, and D98/HR-1 cells. This enzyme was also characterized by its unique electrophoretic mobility and immunospecificity (*Clough* 1979; *Cheng* et al. 1980). Exposure of Raji cells and P3HR-1 cells to TPA and *n*-butyrate induced the same enzyme (*Kawanishi* and *Ito* 1983). This enzymatic activity was specifically inhibited by sera from NPC patients with anti-EA antibody (*Cheng* et al. 1980; *Kawanishi* and *Ito* 1983). No such enzymatic activity was detectable in genome-negative BJAB cells otherwise treated in the same way by IUdR and/or *n*-butyrate. These data suggest that the lack of induction of this enzyme in nonproducer cells is not responsible for incapability of viral DNA synthesis and that this enzyme could be one of the EA complex.

3.3 Induction of EBV Antigens by Other Tumor Promoters

Plants other than *Croton tiglium* contain different types of diterpene esters with tumor promoting activity in mouse skin. *Hecker* (1978) classified these into three classes – tigliane, ingenane, and daphnane types. For their structures and tumor promoting activities, readers are referred to the review by *Hecker* (1978).

Zur Hausen et al. (1979) studied EBV induction by non-TPA deterpene derivatives and anthralin. These structurally related phorbol esters also exert a number of effects on the phenotype of mammalian cell culture, in addition to their promoting activities on skin cancer in vivo like TPA (*Hecker* 1978). Phorbol-12, 13-didecanoate (PDD), a potent tumor promoter comparable to TPA, maximally induced EA in Raji cells at molarities ranging between 5×10^{-10} and 1×10^{-8}. 4-O-Methyl-TPA, marginally active as a promoter, is a relatively weak inducer in Raji cells (two orders of magnitude less active than TPA). Nonpromoters (4-PDD, phorbol, and ingenol) did not induce EA production. Exceptions to this correlation are 12-deoxyphorbol-13-decatrienoate and resiniferatoxin, both nonpromoters, but nonetheless irritants of considerable activity, which induced viral antigen synthesis, and anthralin, a promoter of medium activity, which did not induce viral antigen production. Interestingly, the percentage of induced cells remained constant over a wide range of concentrations with the potent inducers and did not differ from that obtained with the weak inducers, when the latter were applied at their optimal concentrations. Thus, bearing the exceptions in mind, there seems to be, at least in the series of deterpene derivatives, some correlation between promoting activity and virus induction. There seems to be no correlation at all between irritant activity and virus induction.

Recently, several classes of new tumor promoters which have no structural relationship with deterpene esters were discovered by the *Fujiki* and *Sugimura* group (*Fujiki* et al. 1981, 1982). These naturally occurring tumor promoters were divided into three groups: 1) teleocidin, dihydroteleocidin, lyngbyatoxin A; 2) aplysiatoxin, bromoaplysiatoxin, debromoaplysiatoxin, oscillatoxin; 3) palytoxin. Teleocidin and lyngbyatoxin A are indole alkaloids isolated from *Streptomyces mediocidicus* and the blue-green alga, *Lyngbya majuscula,* respectively. Teleocidin has almost the same biological effects as TPA and is effective at similar concentrations. Teleocidin induces ODC in mouse skin, adhesion of human promyelocytic leukemia cells (*Fujiki* et al. 1979), aggregation of human lymphoblastoid cells (*Hoshino* et al. 1980), inhibition of terminal differentiation of Friend erythroleukemia cells (*Fujiki* et al. 1979), and an increase in the uptake of 2-deoxyglucose and prostaglandin synthesis (*Umezawa* et al. 1981). Application of teleocidin and lyngbyatoxin A caused the efficient induction of EA and/or VCA in Raji and P3HR-1 cells (*H. Yamamoto* et al. 1981b; *Eliasson* et al. 1983a), their effects being maximal at a concentration of 5–12.5 ng/ml. Teleocidin also induced infectious EBV from P3HR-1 cells. The effects of both chemicals on EBV antigen induction were further enhanced by the addition of *n*-butyrate, but not by TPA. All of these data demonstrate that teleocidin and lyngbyatoxin A are very similar to TPA in inducing EBV antigens and that both types of chemicals exert their effects by the same mechanisms. Aplysiatoxin and debromoaplysiatoxin are polyacetates isolated from the blue-green alga, *Lyngbya majuscula.* The potent tumor promoter, aplysiatoxin, induced maximal synthesis of EA at the concentration of 5–10 ng/ml, whereas the weak promoter, debromoaplysiatoxin, required a concentration of 250 ng/ml to achieve maximal induction. Superinduction of EA was also observed when these promoters were used in combination with *n*-butyrate. No data are available for the inducibility of EA by oscillatoxin. All the compounds in these two classes inhibited the binding of ^{3}H-TPA to a mouse particulate fraction competitively (*Fujiki* et al. 1983). This suggests that all these tumor promoters can interact with the same receptor in target cells. *Weinstein* (1983) proposed a stereochemical model to explain the interaction of these amphipathic molecules, and of teleocidin and aplysiatoxin, with this receptor system. A third type of promoter, palytoxin, was positive for the irritant test, while in the other tests, induction of ODC, adhesion of HL-60 cells, and inhibition of the specific binding of ^{3}H-TPA to a mouse particulate fraction, the results were all negative. Whether this chemical induces persisting EBV genomes in lymphoblastoid cell lines remains to be determined.

A serum factor which is able to induce synthesis of EBV antigens has been reported by *Bauer* et al. (1982c). It was present in the sera of vertebrates of all five classes tested, including man. Several BL-derived cell lines (Raji, Daudi, EB3, GOR, and Jijoye) and EBV-transformed marmoset cell lines responded to the factor, while human LCLs were not induced by the serum factor. One of the most striking features of this factor is the necessity for treatment by pH shock to gain maximal activity. Although the factor was purified to homogeneity and characterized biochemically (*Bauer* et al. 1982b, 1982c), there are no data available on whether this factor shares other properties with tumor promoters besides inducing activity. It is conceivable that this factor may play an important role in the activation of latent viral information under physiological conditions.

EGF has been shown to act as a potent mitogen (*Hollenberg* and *Cuatrecasas* 1973; *Gregory* 1975) and shares a number of biological effects with TPA (*Lee* and *Weinstein* 1968a; *Stastny* and *Cohen* 1970; *Hamburger* et al. 1981). Furthermore, recent studies have shown that EGF enhanced the transformation steps initiated by X-ray irradiation, UV light, Ad5 (*Fisher* et al. 1981), polyoma virus (*Seif* 1980), and Kirsten sarcoma virus (*Harrison* and *Auersperg* 1981). EGF, however, did not induce EA in Raji cells (*Yamamoto*, unpublished data).

3.4 Induction of EBV by Tumor Initiators and Intercalating Agents

Zur Hausen et al. (1979) studied the effects on EA induction in Raji cells elicited by initiators of the polycyclic aromatic hydrocarbon type of mouse skin tumor and by the ultimate liver carcinogen N-acetoxy-N-2-acetylaminofluorene. All the initiators tested [7, 12-dimethylbenz(a)anthracene, benzo(a)pyrene, 1:2,5:6-dibenzathracene, 20-methylcholanthrene, anthracene, 4-nitroquinoline-1-oxide, and N-acetoxy-N-2-acetylaminofluorene] failed to induce EA at biologically active concentrations. Interestingly, another type of mutagen, N-methyl-N'nitro-nitrosoguanidine (MNNG) (0.5–1 μg/ml) induced EA and VCA in B95-8 and QIMR-WIL cells (*H. Yamamoto* et al., unpublished data). *Lin* et al. (1983b) reported a similar observation that EBV genome copies per cell in producer cell lines increased threefold, as determined by cRNA-DNA hybridization, after treatment with 0.5–2 μg/ml MNNG. This chemical failed to have the same effect in nonproducer Raji and NC-37 cells. Concomitant treatment of P3HR-1 cells with MNNG and TPA showed a further ninefold increase in the number of EBV genome copies per cell. When Raji cells were pretreated with MNNG and then superinfected with P3HR-1 virus, a 35% increase in EBV DNA replication was noted, using density centrifugation analysis. In contrast, treatment of virus with MNNG prior to infection of Raji cells resulted in a marked reduction in infectivity, indicating that intact viral genome may be needed for infection (*Lin* et al. 1983b).

Ernberg and *Moar* (1981) showed that mitomycin C (MMC, 0.5 μg/ml), an alkylating chemical like MNNG, induced EA and VCA in four EBV producer cell lines (P3HR-1, Daudi, Maku, Onesmas), but not in eight nonproducer cell lines, including the Raji line. Parallel to the increase in viral antigens, viral DNA synthesis in the producer lines was increased about twofold. Under these conditions, host cellular DNA synthesis was inhibited by 99% in the presence of MMC.

An unusual cell specificity for efficient EA induction was seen after treatment with various intercalating chemicals, including ethidium bromide (ETB), actinomycin D (AMD), acridine orange (AO), methyl green (MG), and trioxsalen (TX) (*Yamamoto* and *zur Hausen* 1981). All these agents, except MG and TX, effectively induced EA in B95-8 cells only, but not in ten additional cell lines, such as P3HR-1, QIMR-WIL, Raji, or NC-37. Growth of B95-8 cells was completely inhibited when the same concentration of these drugs was used for virus induction. These data suggest that the control mechanism(s) of EBV genome expression in B95-8 is different from that of the other lymphoblastoid cells.

After brief treatment with the intercalating agent AMD, RNA synthesis in B95-8 cells rapidly decreased. This inhibition was irreversible, and there was a reverse correlation between inhibition of RNA synthesis and induction of EA (*Yamamoto,* unpublished data). Although we have no evidence that intercalating chemicals induce EA in B95-8 cells without the stimulation of virus-specific mRNA, it is conceivable that the dyes act on a post-transcriptional step of EA protein synthesis. *Tomkins* et al. (1972) presented evidence that in numerous systems inhibitors of macromolecular synthesis paradoxically induced various enzymes and proteins and indicated that such a mechanism for protein induction was more common than previously realized. This mechanism may also be applicable to EA induction by MNNG and MMC, since strong inhibition of cell growth is always associated with effective induction in these systems too.

3.5 Mechanism of EBV Induction by TPA

3.5.1 Inhibition of Induction

Treatment of cells with tumor promoters results in a fast and reproducible increase in the activity of ODC to levels up to 250 times higher than normal activity. It has been reported that this ODC induction by TPA is effectively inhibited by pretreating the cells with retinoic acid (*Verma* and *Boutwell* 1977; *Kensler* et al. 1978). Moreover, retinoic acid has been reported to act as an inhibitor in experimental two-stage carcinogenesis models (*Shamberger* 1971). On the basis of these facts, the effect of retinoic acid on TPA induction of EBV has been studied (*Yamamoto* et al.

1979). Concomitant treatment of Raji cells with TPA and retinoic acid resulted in effective inhibition of EA induction by TPA. At 10^{-5} Mol, more than 90% of EA induction by TPA was suppressed, and some inhibition was still obvious at 10^{-8} Mol. Even pretreatment of the cells followed by washing and addition of TPA resulted in significant inhibition of EA induction by this chemical.

The specificity of the inhibitory effect of retinoic acid on EBV antigen induction by tumor promoters was then investigated. Raji cells induced by IUdR or anti-IgM-antiserum to EA synthesis were treated simultaneously with varying concentrations of retinoic acid. Although IUdR induction of EA seemed to be less strongly inhibited than TPA induction, some inhibition was obvious over a relatively wide range of retinoic acid concentration. EA induction by anti-IgM is clearly inhibited by retinoic acid. Similarly, EA induction by the intercalating dyes in B95-8 cells was efficiently inhibited by this drug (*Yamamoto* and *zur Hausen* 1981).

Further experiments were designed to test the influence of retinoic acid on EA induction by superinfecting Raji cells with EBV derived from P3HR-1 cells. Retinoic acid did not affect the percentage of EA-positive cells, even when applied at 10^{-5} Mol. Similarly, the spontaneous induction rate of P3HR-1 and B95-8 cells, which is approximately 6% and 4% respectively, was not significantly altered by applying retinoic acid at $10^{-7}-10^{-5}$ Mol for 3 and 5 days. However, various retinoic acid derivatives were found to be effective in inhibiting EA induction by TPA (*Yamamoto* et al., unpublished data; *Zeng* et al. 1981). These data show that retinoic acid inhibits viral induction by chemical inducers and anti-IgM antiserum without affecting antigen synthesis after superinfection or affecting spontaneous induction. This suggests that the different inducing reagents activate the persisting genomes indirectly by a common mediator which seems to be inhibited by retinoic acid. EA induction by superinfection and spontaneous induction of viral antigens in tissue culture cells apparently follow different pathways. Such observations are consistent with the demonstration of a correlation between inducibility by the promoters, IUdR and anti-IgM and EBV genome equivalents per cell, whereas the response to superinfection seems to be independent of genome numbers in EBV genome-carrying cells, as shown in the following section.

Several hypotheses have been advanced to explain the mechanisms of the anticarcinogenic activity of retinoic acid (*Lotan* 1980). It is at present difficult to say whether induction of viral antigens and tumor promotion by TPA are related. However, the primary action of TPA appears to be at the cell surface membrane, and various changes in cell membrane structures occur without requiring de novo RNA and protein synthesis. These changes presumably account for several subsequent effects of TPA. The ability of retinoids to inhibit induction of EBV by TPA could be due to the opposite effects of the retinoids at the cell membrane level. It is also possible that the inhibition of viral antigen induction by TPA takes place through its labilization of lysosomal membranes, leading to increased lysosomal enzyme activity, destruction of proteins, and leakage of intracellular components (*Shamberger* 1971). These changes induced by retinoic acid could also explain the effective inhibition of EBV induction by other inducers, such as IUdR, *n*-butyrate, antihuman IgM, intercalating chemicals, and serum factor.

Similar results were obtained in experiments with L-canavanine, an amino acid analog of arginine (*Yamamoto* et al. 1980a). This uncommon amino acid is found in a number of bacteria, plants, and viruses (*Shieve* and *Skinner* 1963; *Cushing* and *Morgan* 1952; *Pearson* et al. 1952). EA induction of NC-37 cells by TPA, IUdR, antiserum to human IgM, and *n*-butyrate was clearly inhibited by this treatment. Application of 0.3 m*M* canavanine resulted in more than 95% inhibition of EA induction by TPA. Under these conditions, cell growth and incorporation of radiolabeled amino acids into an acid-insoluble fraction was significantly impaired. On the other hand, the analysis revealed that after infection of BJAB and NC-37 cells with P3HR-1 EBV, synthesis of EA was not affected by canavanine in a concentration of up to 8.4 mMol. The synthesis of EBNA and of VCA was significantly inhibited at concentrations higher than 2.8 mMol. In contrast, spontaneous induction of EA in P3HR-1 cells was not affected by canavanine. The data obtained with canavanine again demonstrate that antigen induction by promoters and induction by exogenous infection follow different pathways. However, there is one interesting difference between the inhibition of EA induction by canavanine and inhibition by retinoic acid. In contrast to the latter, treatment of the cells with canavanine prior to induction does not interfere with EA synthesis. Retinoic acid is still active in inhibition of EA synthesis when the cells are treated overnight prior to removal of the drug. Retinoic acid inhibits EA induction in a relatively wide range of concentrations which do not affect growth of the cells. On the other hand, canavanine reduces EA induction at concentrations which also inhibit cell growth and protein synthesis. As will be shown in the following section, other protein inhibitors, such as cycloheximide (CH) and puromycin, also inhibit EA induction by inducers without affecting EBV antigen synthesis by superinfection (*Yamamoto* et al. 1980b). This suggests that these groups of inhibitors might inhibit EA induction through their inhibitory effects on cellular protein synthesis. There is no indication that retinoic acid inhibits cellular protein synthesis at concentrations which eliminate EA induction by inducers (*Yamamoto,* unpublished data). Further studies are necessary to gain more information on the different modes of inhibition by canavanine and retinoic acid.

It is uncertain whether the inhibition of ODC activity by retinoic acid has any direct relationship to the inhibition of viral induction by retinoic acid. Preliminary data indicate that neither putrescine, which is thought to interact with ODC by feedback inhibition, nor colchicine, which is also reported to inhibit TPA-induced ODC activity, reduce the number of TPA-induced Raji cells (*zur Hausen,* unpublished data).

It has been shown that proteases are involved in the process of carcinogenesis and proliferation of cancer cells (*Unkeless* et al. 1973; *Ossowski* et al. 1973; *Quigley* 1979). Furthermore, various protease inhibitors have been shown to inhibit carcinogenesis in some systems (*Troll* et al. 1970). It has also been demonstrated that the proteolytic cleavage of lambda phage repressor participates in prophage induction (*Roberts* and *Roberts* 1975; *Meyn* et al. 1977). EBV antigen induction by IUdR was significantly suppressed by various microbial and synthetic protease inhibitors (*Morigaki* et al. 1981), whereas no clear answer was obtained when the same type of experiments were carried out with TPA (*Yamamoto,* unpublished data; *Sugawara,* unpublished data). These data on protease effects on EBV induction by TPA show some differences from data obtained in endogenous murine xenotropic virus systems (*Hellman* and *Hellman* 1981), where TPA and protease treatment led to a significant decrease in virus production. As discussed below, time-course experiments with TPA revealed that induction requires a longer time to reach a plateau (about 3 or 5 days) (*Yamamoto* et al. 1980b) than in retroviral induction (1 or 2 days). This suggests that TPA action on retroviral induction takes place in a more direct way than in the EBV system and points to the involvement of proteolytic mechanisms in the induction process of retroviruses.

Sundar et al. (1981) reported that corticosteroids known to be anti-inflammatory drugs inhibited EBV antigen synthesis by TPA. In contrast, *Bauer* (1983b) found that steroid hormones did not inhibit EA induction by TPA, but did induce EA and VCA in a particular producer cell line, Daudi. This induction of EBV antigens was specifically inhibited by TPA, which exhibited no inducing capacity when applied alone in this system. *Magrath* et al. (1979) also reported that corticosteroids enhanced, rather than inhibited, EBV replication in producer cell line P3HR-1, induced by a combination of TPA and low temperature.

3.5.2 Cloning Experiments

Since there is a possibility that cultures derived from single cells might differ in their EA synthesis response, Raji cells were cloned in semisolid agar and individual colonies tested. A very different response to EA induction after TPA treatment was observed in the cloned Raji cells (*Bister* et al. 1979). Cell smears from different clones were prepared 5 days after addition of the tumor promoter. They were analyzed for EA synthesis by indirect immunofluorescence, then the percentage of cells expressing EA in each of 24 clones was determined. Synthesis of EA varied among the different clones, most of them exhibiting between 5% and 10% EA-positive cells. This is a similar range to the percentage of EA-positive cells in uncloned Raji cells (6%) after treatment with TPA. Two clones, nos. 11 and 7, contained, respectively, a very low (0.5%) and a very high (25%) number of EA-positive cells. These two clones were tested repeatedly for inducibility of EA synthesis by TPA. In several experiments, cells of clone 11 could not be activated at all, whereas clone 7 occasionally showed up to 40% EA-positive cells.

These clones showed no differences with regard to EBNA expression under these conditions, with or without TPA treatment. Also, cells of clones 7 and 11 did not show any morphological differences and proliferated with essentially identical growth rates in fluid medium. It should, however, be noted that cells of clone 7 have a higher rate of spontaneous EA expression than do cells from clone 11. This may indicate that inducibility by TPA in general depends on the rate of spontaneous viral expression in both producer and nonproducer cell lines.

The detailed analysis of two clones with very different inducibility of EA synthesis showed that the same relative differences in inducibility are observed when the tumor promoters IUdR or antihuman IgM serum are used as activating agents. However, the number of cells activated by EBV superinfection was nearly identical in all cultures tested. It appears that there is no linkage between susceptibility to superinfection and inducibility of the resident genome. This is at variance with a previous study (*Klein* and *Dombos* 1973). However, in this study, different cell lines were compared for their response to superinfection and IUdR, and the proposed linkage was not observed in all lines tested. The lack of linkage between susceptibility to superinfection and activation by chemicals or antiserum observed here suggests that different mechanisms exist for EA activation.

The different responses of cloned Raji cells to nonviral, chemical induction of EA synthesis suggested that the amount or state of EBV-specific DNA in these clones may vary. To test this possibility, the relative number of EBV genome equivalents per cellular genome of the different clones was analyzed by DNA reassociation kinetics.

Tritium-labeled EBV DNA was reassociated in the presence of total cellular DNA from each of the clones. The results demonstrated that the cells of clone no. 7 contained a significantly higher amount of EBV-specific DNA than uncloned Raji cells or cells from clone no. 11. In repeated experiments, the average number of EBV genome equivalents per diploid cellular genome was found to be 110–120 for clone no. 7, 50–60 for uncloned Raji cells, and 40–50 for cone no. 11. These results suggest that: 1) There is a correlation between the susceptibility of chemical induction of EA synthesis and the EBV DNA copy number in different clones of nonproducer Raji cells; 2) individual cells of a given lymphoblastoid line, such as Raji, shown to contain a stable average copy number of 52 EBV genome equivalents per cell (*Nonoyama* and *Pagano* 1973) contain different numbers of viral genomes. EA induction by tumor promoters, base analogs, or anti-IgM serum might be possible only in cells with a certain minimum number of EBV genome equivalents, the fraction of these cells being related to the average copy number per cell of the cultures tested. On the other hand, EA activation by superinfection is obviously possible in every cell expressing EBNA; EA induction is only related to the multiplicity of infection, perhaps by a mechanism of complementation between the input and the resident viral genomes (*zur Hausen* and *Fresen* 1977). A possible relationship between EA inducibility and the number of EBV genome equivalents per cell has also been proposed in studies comparing different lines with regard to their inducibility by IUdR and content of EBV genome equivalents (*Pritchett* et al. 1976; *Moar* et al. 1978).

3.5.3 Superinduction of EBV Antigens by Different Combinations of Inducers

Another approach to studying the control of persisting EBV genomes is represented by superinduction experiments. As shown previously, various chemicals besides TPA, such as IUdR, antihuman IgM, *n*-butyrate, and a recently described serum factor are effective inducers of EBV synthesis. However, even at optimal concentration of each drug, only a relatively small percentage of cells in the total population is induced to synthesize the viral antigen. It has been shown that certain combinations of inducers greatly expand the reactive cell population. The number of induced cells was much higher than the number of cells induced by treatment with individual inducers. For example, TPA plus IUdR (*Yamamoto* et al. 1980b), TPA plus *n*-butyrate (*Lenoir* 1979; *Yamamoto* et al. 1980b), and IUdR plus antibody to IgM (*Tovey* et al. 1978) showed cooperation. In contrast, no cooperation was observed between IUdR and *n*-butyrate. Interestingly, a serum factor could efficiently cooperate with all these inducers (*Bauer* et al. 1982c). *Bauer* (1983a) studied this phenomenon extensively on the basis of quantitative relations between the doses of combined inducers and induction. These studies showed two significant changes associated with superinduction: First, the plateau values for inducible cells were higher, according to the concentration of the second inducer added to saturated concentrations of the first inducer; second, the induction response to the dose of the first inducer was altered to a much more sensitive range by the second.

These data suggest that all inducers except IUdR/*n*-butyrate induce EBV antigens by different mechanisms. As no synergistic effect was observed between IUdR and *n*-butyrate, they were believed to induce identical effects and were categorized in the same group of inducers. In this regard, TPA and the newly discovered promoters teleocidin and lyngbyatoxin A may be very similar. Since the effects of teleocidin and lyngbyatoxin A on EBV antigen induction are further enhanced by the addition of *n*-butyrate, but not by TPA, as described in *H. Yamamoto* et al. (1981b), both chemicals should be categorized in the TPA group. The fact that a serum factor can cooperate with any of the four chemical inducers may indicate that this serum factor acts at a central step within the induction process. The existence of a central and common step for induction is further supported by the sensitivity of all inducers to retinoic acid and when used in combination (*Bauer* 1983a).

3.5.4 Selective Stimulation of Chromosomal Protein Synthesis by TPA

Very recently, *Lin* et al. (1983) studied various physiological conditions in order to establish the optimal conditions for TPA induction. Interestingly, as the time of TPA addition to the culture of P3HR-1 cells became progressively later, there was an increasing induction effect. The greatest response to TPA induction occurred in nonproliferating cells (15-fold greater than the control level), as opposed to cells growing exponentially (sixfold greater than the control level). This result was also supported by the fact that treatment of resting P3HR-1 cells with TPA resulted in a threefold further increase in biologically active virus (determined by superinfection of Raji cells), compared with virus from exponentially growing cells treated with TPA. These findings represent a sharp contrast to IUdR induction of EBV, in which DNA synthesis is a prerequisite for induction (*Hampar* et al. 1972).

Under nongrowing cell conditions, an overall reduction in total cellular protein synthesis occurred. However, a selective increase in chromosomal protein synthesis accompanied that in EBV genomes in P3HR-1 cells. Analysis of chromosomal proteins synthesized in P3HR-1 cells after treatment with TPA revealed that six distinct polypeptides of 145K, 140K, 135K, 110K, 85K, and 55K were selectively induced. All these polypeptides appeared to be EBV specific since: 1) By the use of immunoprecipitation techniques, similar polypeptides were shown to be synthesized after superinfection of nonproducer cells with P3HR-1 EBV (*Bayliss* and *Nonoyama* 1978; *Mueller-Lantzsch* et al. 1980a),.or after induction of the cells with chemical inducers; 2) such polypeptides were not synthesized by EBV-negative lymphoblastoid cells after TPA treatment (*Mueller-Lantzsch* et al. 1979, 1980b; *Strnad* et al. 1979; *Thorley-Lawson* and *Edson* 1979).

Regarding the enhancement of activation of latent EBV genomes in nonproliferating cells, it was suggested that endogenous viral functions may take over under such conditions and proceed from latency into a replicative cycle, probably due to the difference between viral and cellular DNAse (*Datta* et al. 1980b). It is important to know whether some of these chromosomal proteins play a role in the regulation of expression of latent EBV genomes.

3.5.5 Effect of Inhibitors of Macromolecular Synthesis

Phorbol diesters stimulate incorporation of nucleic acid precursors into various types of cells (*Fusenig* and *Samsel* 1978; *O'Brien* et al. 1979). *Yuspa* et al. (1976) reported that DNA stimulation in cells correlated with tumor-promoting activity. It was, therefore, of great interest to investigate the role of stimulation of cellular DNA synthesis in viral induction (*Yamamoto* and *zur Hausen* 1980). Simultaneous treatment of the P3HR-1 cells with an inhibitor of DNA synthesis, Ara-C, and TPA did not block EA or VCA induction, and VCA-positive cells synthesized DNA even in the presence of the inhibitor. This experiment showed that simultaneous treatment of the cells with TPA and Ara-C did not completely suppress DNA synthesis of the cells, possibly due to the DNA-stimulating effect of TPA. However, pretreatment of the cells with Ara-C prevented DNA synthesis completely. Even under these conditions, TPA induced EA, whereas VCA induction was efficiently suppressed. Similar results were obtained in the expression of EA with nonproducer cells.

Essentially the same data were presented by *Ooka* and *Calender* (1980), who used Ara T. The expression of VCA by the combination of TPA and *n*-butyrate was inhibited in the presence of Ara T, whereas no effect on EA production was observed. The activation of P3HR-1 thymidine kinase minus cells by the combined treatment of TPA and *n*-butyrate induced high levels of ^{3}H-thymidine incorporation into viral DNA, which was entirely inhibited by the analog, as demonstrated by CsCl gradient analyses.

Acyclovir[9-(2-hydroxyethoxymethyl)guanine] is a potent anti-herpesvirus drug which inhibits EBV DNA replication and viral antigen synthesis. *Lin* et al. (1981) studied the effect of this chemical, comparing it with that of retinoic acid on TPA induction of EBV. Acyclovir (100 μMol) completely inhibited the stimulation of ^{3}H-thymidine incorporation into viral DNA in both TPA-treated and nontreated P3HR-1 cells. In contrast, retinoic acid strongly inhibited EBV DNA replication in TPA-treated cultures exclusively, without interfering with spontaneous viral DNA replication. This discriminatory effect of acyclovir and retinoic acid was further supported by cRNA-DNA hybridization with EBV-specific cRNA. Similarly, *Ooka* (1983) showed that two other types of anti-herpes-virus drug, phosphonoacetic acid and phosphonoformate, inhibited viral DNA synthesis and VCA synthesis in P3HR-1 cells after TPA treatment without interfering with EA synthesis.

These studies show that the early steps leading to induction by the tumor promoter TPA are independent of cellular DNA synthesis. Neither abortive induction of nonproducer cells, resulting in EA synthesis only (*zur Hausen* et al. 1978; *Fresen* et al. 1978; *Bister* et al. 1979), nor induction of EA in producer cells by TPA is significantly affected by the inhibition of cellular DNA synthesis. This again contrasts with viral induction by halogenated pyrimidines, which induce persisting viral genomes after incorporation into host cell DNA and thus require cellular DNA synthesis as a prerequisite for induction (*Gerber* 1972; *Klein* and *Dombos* 1973).

So far, there is no experimental evidence of the cell-cycle-dependent control mechanism of EBV expression, since synchronization of Raji cells did not result in complete EA activation in all cells via some form of induction system, including chemicals and serum factors. Furthermore, even prolonged treatment of Raji cells with TPA did not result in EA activation in all cells of the culture tested, although the continued growth of the cells suggested that most of them traversed a complete cell cycle. The data obtained with DNA inhibitors may also support the view of cell-cycle-independent regulation of EBV expression.

The next question is whether TPA directly activates transcription into mRNA of the gene loci which code for the EBV polypeptides. It has been shown, for example, that ecdysone, the steroidal molting hormone, acts directly on the chromosome, since puffing of certain loci can occur even in the absence of protein synthesis. We used AMD as the main tool for investigating RNA involvement in the induction of EBV by TPA (*Yamamoto* et al. 1980b). We used 0.25 μg/ml AMD to inhibit RNA transcription; it was added to NC37 cells either simultaneously with TPA or 1, 2, 3, or 4 days after TPA addition. Samples were collected after different incubation times and examined for the percentage of EA-producing cells by indirect immunofluorescence. The results indicated that concomitant addition of inhibitor and TPA leads to complete inhibition of EA induction. However, if the inhibitor was added at intervals after prior induction, some EA induction occurred, correlating with the inhibitor-free time period. AMD added 3 days after induction had no effect on subsequent EA expression. This suggests that sufficient message accumulated within 2 days, but that EA expression is not influenced by AMD addition after this period.

The next series of experiments was designed to study whether EA induction by TPA, as well as by superinfection, depends on prior synthesis of non-EA proteins. Cells were incubated with TPA and CH (50 μg/ml) for various periods of time (2, 3, and 5 days), washed thoroughly three times with phosphate-buffered saline (PBS), and then further incubated. Previous studies had revealed that EA induction by TPA resulted in the appearance of the first antigen-positive cells about 1 day after addition of the promoter. In such cultures, the highest percentage of induced cells was obtained about 5 days after induction. Therefore, it was of interest to determine whether the initial inhibition of protein synthesis for various periods of time by CH, followed by the release of this block, would permit the quick appearance of EA-positive cells. The results showed that the kinetics of EA induction after CH treatment and removal of the inhibitor follow a very similar pattern to those of non-CH-treated cultures, irrespective of the initial incubation time with CH.

In a second set of experiments, NC37 cells were treated concomitantly with TPA and CH (50 μg/ml) for 2 days, washed three times with PBS, and then divided into four portions. Subsequently, they were treated with and without AMD (0.25 μg/ml) and with and without TPA. The culture which received inhibitor-free medium showed a gradual increase in EA-positive cells. The addition of AMD, however, completely prevented the appearance of EA-positive cells, irrespective of the presence or absence of TPA. Interestingly, the addition of fresh TPA did not cause any further increase in the percentage of EA-positive cells. These data suggest that TPA induction is indirect and that synthesis of a specific protein, or proteins, precedes induction of EA.

Additional experiments were carried out without using TPA but superinfecting NC37 and EBV genome-negative BJAB cells with P3HR-1 EBV. After 2 days, nonabsorbed virus and inhibitor were removed by washing, and this was followed by AMD treatment, as described above. In these tests, a prompt increase in EA-positive cells occurred after removal of CH, irrespective of the presence or absence of AMD. Only 3 h after CH removal, definite induction of EA was visible, reaching a maximum 1 day after addition of AMD. Both types of cells provided the same data, although there was a significant difference in the percentage of induced cells, as observed in earlier experiments (*zur Hausen* and *Fresen* 1977). This strongly suggests that the initial CH treatment did not block transcription of the superinfecting EBV genomes and supports the theory that induction of endogenous EBV genomes is regulated differently to that of superinfecting EBV DNA.

3.5.6 Mechanism of Induction

The question arises as to whether the inducer TPA produces an increase in the number of cells capable of making EA, or whether EA synthesis occurs in all of the cell population at undetectable levels, even in the absence of an inducer. Alternatively, TPA may induce an undifferentiated precursor cell to transform into a more mature cell. This assumption is especially relevant in view of recent observations that TPA is a highly potent chemical that alters the differentiation program of cells in a variety of systems, as described in a preceding section. Indeed, TPA treatment of EBV-positive lymphoblasts resulted in increased adhesiveness, aggregation, and enlargement of the cells, associated with the polyploidization (*zur Hausen* et al. 1978). *Anisimova* et al. (1982) indicated the involvement of cell differentiation in EBV induction in Raji cells by *n*-butyrate. By means of electron microscopy, they found apparent signs of differentiation in drug-treated Raji cells in the direction of plasma cells. Butyrate, however, induced no differentiation-related change in P3HR-1 cells. It is also not clear whether EA is favorably induced in differentiated Raji cells. In the case of TPA induction, I believe mechanisms other than differentiation mechanisms may be involved. For example: Induction can occur in either dividing or resting cells, as shown in the experiments with inhibitors of macromolecular synthesis (*Yamamoto* and *zur Hausen* 1980; *Ooka* and *Calender* 1980), and with nonproliferating or proliferating states of the same cells (*Lin* et al. 1983a). Induction seems to depend on the copy number of EBV genomes (*Bister* et al. 1979). TPA induction, therefore, does not appear to depend on cellular differentiation.

Recent studies have suggested that DNA methylation plays an important role in gene expression of higher organisms at the transcriptional level. This assumption has come from two basic observations: 1) An inverse correlation has been observed between the level of DNA methylation and the level of gene expression of mRNA in several viral systems (*Sutter* and *Doerfler* 1978; *Niwa* and *Sugahara* 1981; *Youssoufian* et al. 1982) and in a number of eukaryotic genes (see reviews, *Razin* and *Riggs* 1980; *Ehrlich* and *Wang* 1981); 2) induction of muscle cell differentiation is caused by the incorporation of the cytidine analog 5-azacytidine into the DNA of mouse cells (*Jones* and *Taylor* 1980).

Several groups have studied the possible role of DNA methylation in the EBV system (*Diala* and *Hoffman* 1983; *Larocca* and *Clough* 1982). *Kintner* and *Sugden* (1981) found that two segments of episomal EBV DNA of a newly established nonproducer cell line showed increasing methylation with time. *Perlmann* et al. (1982), using the cloned EcoR1-J fragmen of the B95-8 strain of EBV as a probe, reported that EBV virion DNA consisted of a mixture of both methylated and nonmethylated DNAs. They did not find any difference in the methylation pattern in this region between transforming and nontransforming viruses. Employing the isoschizomers Hpa2 and Mspl and the cloned EcoR1-J fragment as a probe, *Saemundsen* et al. (1983) studied whether TPA and/or *n*-butyrate caused differences in the methylation pattern after treatment. They found that intracellular EBV DNA was methylated in and around the EcoR1-J fragment in both producer and nonproducer cell lines, and this methylation was more pronounced in the latter cells than the former. However, no obvious differences were observed in the methylation pattern at this site before and after induction.

Ben-Sasson and *Klein* (1983) tested the effect of 5-azacytidine in the EBV system. Results showed that only a short exposure to the chemical caused induction of EA in a substantial number of the cells in all six EBV producer lines (P3HR-1, Jijoye, Daudi, and three hybrid cell lines). However, EBV nonproducer lines (Raji, Namalwa, and two hybrid cell lines) did not respond to drug treatment. These data suggest that, at least in producer cell lines, the changes in EBV genome expression may be related to the state of DNA methylation. In EBV-infected human B lymphocytes, most of the EBV genomes exist episomally (*Nonoyama* and *Pagano* 1972), and this episomal EBV DNA is responsible for the production of EBNA (*Ernberg* et al. 1977), which is essential for the maintenance of the transformed state of the lymphocytes by EBV. It is possible that producer cells may, in addition, contain EBV genomes integrated at sites where the flanking host DNA sequences might trigger the expression of EA genes after activation. In the case of nonproducer cells, the location of EBV genomes in cellular DNA may not be suitable for bringing about gene expression after chemical treatment. It is also possible that some deletion of EBV genomes in nonproducer cells is an important factor in this discrepancy.

Recently, *Eliasson* et al. (1983b) investigated whether induction of EBV by chemical inducers was due to demethylation of regions that control transcriptional activity. Demethylation was detected in a producer line, P3HR-1, but not in a nonproducer Raji cell line after concomitant treatment with TPA and *n*-butyrate. Demethylations were seen at two Sma sites in the EcoR1-J fragment and Bam H1 A fragment of EBV DNA. These regions contain, respectively, intragenic control regions and sequences coding for a DNA-binding early protein. The latter part of the genome is deleted in Raji EBV DNA. No detectable change in methylation was induced by the application of TPA alone.

Unlike other viral systems, the role of DNA methylation in EBV gene expression is, thus, not clear at this time. Further studies are obviously needed to elucidate this role. This particularly applies to studies involving the mechanism of tumor promoter action and latent and productive infection of EBV genomes.

Lau et al. (1981) first described the effect of TPA on the transcription of EBV genomes in two somatic cells hybrids of human lymphoma and human myeloid leukemia cells, Putko (P3HR-1 plus K562, *Klein* et al. 1980) and Dutko (Daudi plus K562, *Lau* et al. 1981). In these particular cell lines, increased transcription of viral RNA did not occur after the application of TPA, whereas viral RNA transcription increased from 10% to 25% in Putko, and from 20% to more than 40% in Dutko cells after IUdR treatment. However, studies with B95-8 cells (*Hummel* and *Kieff* 1982; *Shin* et al. 1983) and GC68-13, a subclone of P3HR-1 cells (*Weigel* and *Miller* 1983), suggested that TPA indeed stimulated the transcriptional step of EBV genomes. These authors used TPA as a tool for studying the difference in the transcripts produced during the latent and productive states of the cells, rather than as a means of studying the mechanism of TPA action on EBV induction. Therefore, no extensive kinetic studies were done on TPA stimulation of EBV transcription. Nevertheless, these studies have provided very important information on the mechanism of TPA action. *Weigel* and *Miller* (1983) selected 48 h after TPA treatment for the preparation of RNA from the GC 68-13 subclone of P3HR-1 cells, which spontaneously synthesize viral antigen at a low rate; more than 80% of the cells entered the viral replicative cycle after TPA exposure. The results showed that: 1) TPA caused a marked increase in the number of different species of mRNA molecules. At least six regions of the genomes are transcribed during latency (*van Santen* et al. 1981; *Weigel* and *Miller* 1983). Following TPA treatment, 54 mRNAs were identified. 2) TPA caused a marked increase in the abundance of stable viral specific mRNAs. 3) Those RNAs transcribed during latent infection also seem to be produced, but in larger amounts, during viral replication after TPA exposure. These results are essentially in agreement with those obtained by *Hummel* and *Kieff* (1982) and *Shin* et al. (1983), using B95-8 cells. These data are also consistent with previous observations which showed the stimulatory activity of TPA on the transcription of specific genes in various systems (*Arya* 1980; *Gottesman* and *Sobel* 1980; *Soprano* and *Baserga* 1980; *Hoffman-Liebermann* et al. 1981; *Fisher* et al. 1981; *Amtmann* and *Sauer* 1982).

Taking all of these data into account, we speculated that viral induction by phorbol diesters would be effected by a cellular mediator synthesized in an initial response to the induction event. This speculation seemed to be supported by the following observations: 1) Maximal induction by TPA requires rather a long time (5 or 6 days); 2) simultaneous treatment of the cells with TPA and CH blocks antigen expression completely, but after removal of the inhibitor there is a kinetic pattern of EA induction similar to that observed in non-CH-treated cells, irrespective of the initial incubation period with CH; 3) in Raji cells treated simultaneously with TPA and CH for 2 days, subsequently washed, and treated thereafter with AMD, no EA was detected. These results seemed to support the view that a specific cellular mRNA has to accumulate during CH treatment of TPA-induced cells, which, after translation and upon interaction with persisting

EBV DNA, leads to the stimulation of transcription of virus-specific mRNA. Such cellular protein must either be a positive regulator, or a negative one, which inhibits the production of cellular repressor that restricts the expression of latent EBV genomes.

Unfortunately, however, additional experiments did not support this interpretation. Treatment of EBV-negative lymphoblastoid B or T lines and various strains of human fibroblasts with TPA, followed by fusion of these cells with genome-harboring Raji cells under conditions of CH and AMD treatment, or treatment of the fused cells with retinoic acid, did not result in significant induction, although appropriate controls without inhibitors were clearly positive (*Yamamoto* et al., unpublished data).

Despite the possibility of alternative interpretations, such as an exclusive nuclear localization of the postulated cellular mediator, and a possible failure of nuclear fusion, the following interpretation appears more plausible: Under conditions of latent infection, EBV, and possibly other oncogenic herpes viruses, code for their own repressors. In this regard, it is very important to note the occurrence of a selective increase in EBV-specific chromosomal protein synthesis after treatment of nongrowing cells with TPA (*Lin* et al. 1983a). Some of these proteins might relate to such repressors. TPA and related promoters either block this repressor synthesis or interfere with its function. This, in turn, would enable the viral DNA to initiate its independent transcription and DNA replication.

Recently, DNA strand breakage has been reported as occurring in human leukocytes exposed to TPA, and it may be relevant to the action of TPA as a tumor promoter in animals (*Birnboim* 1982). Since MNNG and EBV are known to cause breakage in chromosomes and fragmentation of DNA (*Strauss* et al. 1974; *Nonoyama* and *Pagano* 1972a), the increase of viral DNA replication in P3HR-1 cells and superinfected Raji cells after MNNG treatment suggests that the effect of TPA may be associated with its capacity to damage the host cell genome (*Lin* et al. 1983b). These breaks could presumably facilitate the integration of viral DNA in a similar fashion, as shown by *Casto* et al. (1979), or impair the integrity of the host chromosomes, leading to the derepression of EBV genomes.

Although TPA-induced chromosomal damage could be another possible mechanism of EBV induction, the fact that the combination of MNNG plus TPA results in a further increase in EBV genome copies per cell suggests that there are some differences in the mechanisms of action of TPA and MNNG.

It should again be emphasized that these are speculations, which, however, contain some testable predictions. If correct, this model could provide a valuable tool for our understanding of the regulation of specific genes in tumor promotion.

3.6 TPA Induction of Additional DNA Viruses

In the initial study by *zur Hausen* et al. (1978), efficient induction of viral antigen was noted in cells transformed by EBV-related Old-World monkey herpes viruses. TPA caused a 2- to 100-fold increase in viral antigen in the two baboon lines 18b and 9b transformed by the herpes virus papio and in the marmoset T-cell line 77/5 carrying the herpes virus saimiri. Virus production was enhanced to some extent in African green monkey-derived B lymphoblasts AGM-2206 containing EBV-like virus and in T lymphoblasts transformed by Marke's disease virus of chickens. Unfortunately, no extensive studies on TPA action in these systems were made thereafter.

In contrast to the persisting oncogenic herpes viruses, substantial difficulties were encountered in demonstrating the effect of TPA with other types of DNA virus. For example, analysis of BK-virus-transformed cells failed to show evidence of viral induction (*zur Hausen,* unpublished data). However, some laboratories have been able to show an increase in viral DNA and antigens in some viral systems after application of TPA. They are summarized as follows. *Krieg* et al. (1981) showed that a single exposure to TPA (55--672 ng/ml) resulted in a 1.4- to 6-fold increase in viral DNA content in an HD strain of nonproducer Vero cells containing stump-tailed macaque virus. These cells contain viral DNA in an episomal state and show a transformed phenotype. In contrast, no viral capsid antigens were expressed, despite TPA treatment. However, TPA does not change the number of viral genome equivalents in productively infected cells containing HD papova virus or in cells containing the papova virus SV40 in an integrated state. The enhancing effect of the drug on viral replication does not, therefore, seem to be a general phenomenon, and the authors suggest that the physical state of the SV40 DNA may either permit or preclude the stimulating effect. These results also suggest the presence of a cellular regulating factor which determines responsiveness to TPA. TPA is also capable of increasing the DNA content in nonpermissive mouse cells and inducing transcription of bovine papilloma virus type 1 (BPV-1) in mouse embryo fibroblast cells, in which viral genomes remain in a nonexpressed episomal state for a few passages after infection (*Amtmann* and *Sauer* 1982).

The addition of TPA (1–1000 ng/ml) simultaneously with Ad5 infection, or up to 8 h after infection, caused a two- to threefold increase in the final yield of infectious virus in HeLa cells (*Fisher* 1983). This was associated with an acceleration of the appearance of early viral antigen, a 72K DNA-binding protein. CPE also appeared earlier in infected cells after TPA treatment. The acceleration of CPE induced by TPA was not blocked by the addition of leupeptin, antipain, or trans-retinoic acid. Analysis of viral mRNA synthesis indicated that TPA stimulated the synthesis of mRNA from early regions of viral genomes. It is, therefore, likely that the acceleration of CPE by TPA results from the overproduction of viral polypeptides (*Fisher* et al. 1982).

3.7 Induction of Retroviruses by TPA

3.7.1 Oncogenes

In 1969, *Toyoshima* and *Vogt* isolated temperature-sensitive (ts) mutants of the Rous sarcoma virus (RSV) B77 strain involved in cell transformation. Analysis of the mutants resulted in the identification of a gene (later called *src*) in the viral genome necessary for the transformation of chicken fibroblasts (*Hanafusa* 1977; *Vogt* 1977). Subsequent studies clarified the presence of different oncogenes from various retroviruses, and about two dozen such transforming genes have so far been identified (*Bishop* and *Varmus* 1982). Some were shown to have an evolutionary relationship to *src*. One of the most striking findings obtained with complementary DNA (cDNA) to viral oncogenes was the discovery of analogous sequences in normal cells from a wide variety of animals, not only mammals but also lower vertebrates such as fish. Thus, the oncogenes preserved in the virus and the cells have been designated *v-onc* (e.g., *v-src*) and *c-onc* (e.g., *c-src*), respectively. Some of the *c-onc* genes were actually proved to be oncogenic after coupling with viral LTRs, or after insertion into retroviral genomes. Thus, it is most likely that *v-onc* is a gene intially transduced from the genome of normal cells in which *c-onc* plays an important role in normal cell growth and differentiation. The *src* gene product is a phosphoprotein called $pp60^{src}$ with a molecular weight of 60 000 and causes cell transformation (*Purchio* et al. 1977; *Levinson* et al. 1978; *Collett* and *Erikson* 1978; *Erikson* et al. 1978; *Brugge* et al. 1978). This protein has the activity of a cyclic adenosine monophosphate (cAMP)-independent protein kinase and phosphorylates tyrosine residues of the target proteins (*Collett* et al. 1978; *Erikson* et al. 1979). As described, the addition of TPA induces transformation mimicry and enhances the expression of transformation markers (*Weinstein* et al. 1978; *Diamond* et al. 1978; *Weinstein* et al. 1979). This mimicry of transformation includes the stimulation of a protease PA synthesis, change in morphology and growth, and several changes in the functions and structures of cell membranes and cytoplasm.

3.7.2 TPA Action on v-src and c-src Kinase Levels

Since the TPA-induced alterations strikingly resemble cells transformed by the RSV *src* gene, it has been hypothesized that TPA might act by increasing the amount or activity of *v-src* or *c-src*. This possibility has recently been tested by different groups. *Goldberg* et al. (1980) reported that after TPA treatment, little or no change was observed in the level of either *c-src* or *v-src* in Schmitt-Ruppin RSV-transformed chick embryo fibroblasts. *Pietropaolo* et al. (1981) also studied the effect of TPA on these kinases in cells transformed by a ts mutant of the avian sarcoma virus (CEF-ts ASV) and compared the known effects of TPA on cell morphology and PA activity. They concluded, as did *Goldberg* et al., that TPA action on cell morphology and PA activity are most likely not the result of any effect on these protein kinases.

3.7.3 Retroviral Induction by TPA

3.7.3.1 MMTV. *Arya* (1980) found that at 10 ng/ml TPA, MMTV stimulation was nearly 20-fold in a persistently infected mouse mammary tumor-derived cell line, Mm5mt/Cl. In this system, dexamethasone has been shown to induce MMTV 10- to 20-fold. The combination of dexamethasone and TPA apparently resulted in 100- to 200-fold enhancement of virus production.

3.7.3.2 Friend Leukemia Virus (FLV). *Colletta* et al. (1980) showed that TPA enhanced the expression of FLV genome in a 745A19 clone of Friend erythroleukemia cells at 16.7 nMol, as demonstrated by a twofold increase in the amount of reverse-transcriptase-containing particles released into the culture fluid and in the levels of virus-specific intracytoplasmic RNA. The greatest effect was also observed as early as 24 h after treatment. The efficient dose of TPA for viral induction also inhibits DMSO-induced Friend cell terminal differentiation. Moreover, the strongest effect of TPA is on the induction of PA. These results suggest that two specific effects of TPA, i.e., a differentiation block and induction of PA, correlate well with the increase in viral expression in the Friend cell system. Similar results were obtained by *Lipp* et al. (1982) using FEM cells established from Evelyne monolayer cells. They also tested the in vivo effect of TPA on spleen focus formation in STV mice infected with the Friend virus complex (Friend murine leukemia virus/Friend spleen-focus-forming virus). Mice treated with TPA (50 ng/g) for 1 week prior to infection showed a fivefold increase in the number of spleen focus compared with untreated control mice.

3.7.3.3 Mason-Pfizer Monkey Virus (MPMV). *Wunderlich* et al. (1982) demonstrated by the reverse transcriptase assay that TPA (5 ng/ml) causes an up to 12-fold stimulation of MPMV synthesis when the virus is growing in human embryonic kidney (HEK) cells. Contrary to infected HEK cells, similar concentrations of TPA induce stimulation of MPMV synthesis in persistently infected cells in the continuous human tumor cell line A204.

3.7.3.4 Endogenous Murine Xenotropic Virus. *Hellmann* and *Hellmann* (1981) showed that TPA induced endogenous murine xenotropic virus from Al-2 cells derived from the BALB/c mouse, as determined by the infectious-center focus-forming assay on permissive normal rat kidney (NRK) cells. Maximal induction occurred after 24 or 48 h when the cells were treated with 80 ng/ml TPA. Under these conditions, TPA induced infectious virus from approximately 0.03% of the cell population. However, the level of induction by TPA was lower than that observed with 25 μg/ml IUdR, which induces approximately 3% to 4% of the cells to release virus. The combination of TPA and IUdR enhanced retroviral induction to a level threefold higher than that seen with optimum IUdR concentration alone. In contrast, *Lipp* et al. (1982) reported that TPA tended to inhibit the induction of endogenous virus production in B-cell mitogen (lipopolysaccharide)-stimulated spleen cell cultures from BALB/c mice, indicating the pleiotropic effect of TPA.

An interesting effect was observed when protease inhibitors were used simultaneously with TPA for endogenous murine viral induction. The results showed that both antipain and leupeptin significantly decreased viral production by TPA. This provides evidence of protease involvement in TPA induction of endogenous retrovirus (*Hellmann* and *Hellmann* 1983).

3.7.3.5 Other Retroviruses. Simian sarcoma virus (SSV)-transformed rat cells (NKR-SSAV) displayed a twofold increase in particle-associated reverse transcriptase after treatment with 4 ng/ml TPA (*Lipp* et al. 1982). *Hoshino* et al. (1983) showed that TPA, PDD, and teleocidin enhanced the production of MLV from k-BALB cells persistently infected with MLV. In this system, the effects of TPA and PDD were not as strong as those of teleocidin.

A newly discovered human retrovirus (*Poiesz* et al. 1980; *Yoshida* et al. 1982; *Yamamoto* et al. 1982) from adult T-cell-leukemia (ATL) patients, ATL virus (ATLV) (*Hinuma* et al. 1981), is inducible to some extent by TPA treatment. About 20 ng/ml TPA caused a two- to threefold increase in ATLV-specific antigens (ATLA), detectable by the indirect immunofluorescence method in MT-l cells. In this system, however, IUdR was found to be more effective than TPA (*Yamamoto*, unpublished data).

4 Enhancement of Viral Transformation by TPA

4.1 EBV System

It has been established that Epstein-Barr virus (EBV) is capable of transforming human leukocytes in liquid medium to LCL growing indefinitely (*Henle* et al. 1967; *Pope* et al. 1968; *Gerber* et al. 1969; *Miller* et al. 1971, 1974; *Chang* et al. 1971; *Leibold* et al. 1975; *Schneider* and *zur Hausen* 1975; *Katsuki* and *Hinuma* 1975; *Yamamoto* et al. 1976). The transforming activity of EBV has been quantitatively assayed by a dilution endpoint method (*Moss* and *Pope* 1972), by a transformed center assay with the use of semiconfluent human fibroblasts as feeder layers (*Robinson* and *Miller* 1975), and by a colony formation assay on soft agarose (*Yamamoto* and *Hinuma* 1976; *Mizuno* et al. 1976).

TPA not only efficiently induced EBV antigens but also significantly enhanced the transformation of normal human leukocytes from either cord blood or peripheral blood by B95-8 EBV (*Yamamoto* and *zur Hausen* 1979). Enhancement of transformation occurred within a narrow dose range (0.002–1.25 ng/ml) of TPA. At the optimum concentration of 0.31 ng/ml, five to eight times more colonies were counted than in cultures without TPA. As described previously (*Yamamoto* and *Hinuma* 1976; *Katsuki* et al. 1977; *Yamamoto* and *Hinuma* 1978), typical "rod-shaped" EBV-transformed colonies appeared in cultures both with and without TPA, but in the TPA-treated cultures, transformed colonies appeared earlier and were larger. TPA has been reported to stimulate T lymphocytes specifically (*Touraine* et al. 1977). However, in this case, enhancement was not due to such an effect, since the colonies in TPA-treated cultures were also EBNA-positive.

It has been reported that TPA is able to enhance the stable transformation of fibroblast cultures previously exposed to various initiators, such as chemical carcinogens, UV light, X-irradiation, or an adenovirus (*Lasne* et al. 1974; *Mondal* et al. 1976; *Mondal* and *Heidelberger* 1976; *Kennedy* et al. 1978; *Fisher* et al. 1979, 1981). The results indicate that "initiated" cells have a qualitatively different response to TPA from that of normal cells. Several cell lines were tested, including an EBV-negative cell line, BJAB, an EBV-positive cell line, P3HR-1, two Raji clones, and a newly established human cord leukocyte line, CBL-1 (about 2 months after establishment). The data show that, except for CBL-1, TPA did not enhance colony formation in these cells. There was even a reduction in the number of colonies (15%–40% less than in the control, and they were also more dispersed in the presence of TPA). However, TPA caused an approximately threefold increase in colony formation in CBL-1 cells. These results did not seem to be consistent with the augmentation of a transformed phenotype by TPA observed in other systems in which enhanced "anchorage independence" was used (*Fisher* 1983).

The effect of TPA on the transformation of leukocytes after infection with EBV was also assessed by end-point dilution. TPA increased transformation efficiency in liquid culture systems. In the presence of 0.31 ng/ml TPA, two independent experiments yielded about five times more TD_{50}/ ml than in the untreated controls (*Yamamoto* and *zur Hausen* 1979).

As is well known, cultures of PBL from EBV-seropositive healthy donors develop into LCL (*Chang* et al. 1971). The frequency of establishment of LCL is lower and the time required for initiation of growth of transformed cells is longer than in cells from IM patients (*Nilsson* 1979). In the latter case, LCL can be routinely established. TPA was also shown to enhance this spontaneous transformation of B cells by EBV from PBL of healthy seropositive adults (*H. Yamamoto* et al. 1981). In the presence of 0.5 ng/ml TPA, the frequency of establishment of LCL increased and the time required for the appearance of transformation decreased. TPA induced a higher percentage of EBNA-positive cells in the peripheral blood when cocultured with cord blood lymphocytes (CBL).

Several possible mechanisms for the enhancement of transformation of human leukocytes by EBV and of spontaneous establishment of LCL from peripheral lymphocytes have been considered. One is a direct promoting effect of TPA on the growth of cells which have been transformed spontaneously in vivo or in vitro by exogenous EBV. In the soft agarose colony formation assay, possible cell-to-cell interaction may be ruled out. The enhancement of transformed colonies by TPA provides evidence supporting this assertion. Other evidence for this view is that the successful establishment of LCL in culture with neutralizing antibody-containing human cord serum (HCS) was obtained in the presence, but not in the absence, of TPA (*H. Yamamoto* et al. 1981a). The fact that TPA enhanced the plating efficiency of newly established cell lines but not that of "old lines" in agarose indicates that TPA may act on the early steps of transformation, rather than inducing nonspecific colony-stimulating factors that are known to promote cell growth in agarose (*MacPherson* 1973).

Another mechanism of TPA action may, however, play a more important role in the enhancement of LCL establishment, i.e., augmentation of active viral production by TPA. The released virus could then transform coresident susceptible B cells in the given lymphocyte preparation. To analyze this putative mode of TPA action, the effect on enhancement of neutralizing antibody to EBV was examined. The results showed that LCL establishment was markedly suppressed in the cultures with human sera with neutralizing activity against EBV (*H. Yamamoto* et al. 1981a).

Additional mechanisms of TPA action have been identified in the course of studies on the involvement of T lymphocytes in the enhancement of transformation (*Harada* et al. 1981). In virus-infected lymphocyte cultures from EBV-seropositive adult donors, in vitro regression of transformation was suppressed by TPA. EBV-specific and -nonspecific cytotoxicity of T cells generated in mixed cultures of peripheral blood lymphocytes from seropositive adults and of autologous LCL cells was markedly lowered by the presence of TPA in the cultures. However, TPA had little effect on the proliferation of T cells in stimulated cultures, and the addition of TPA to the reaction mixture for the cytotoxicity test did not lower the cytotoxicity. These data show that TPA can abrogate T-cell-mediated suppression of EBV-induced transformation.

To summarize, TPA enhances EBV-induced transformation of human leukocytes by: 1) directly affecting some of the transformation steps; 2) inducing active EBV from the virus genome-positive cells, which infects the coresidental target B cells; 3) inhibiting T cells, which are responsible for the regression of EBV transformation.

Teleocidin has been shown to have very similar biological effects to TPA, including EBV induction and enhanced EBV-induced transformation, as described above. Both spontaneous outgrowth of LCL from peripheral lymphocytes of seropositive healthy donors and colony formation of umbilical cord blood leukocytes in soft agar, induced by the B95-8 strain of EBV, were enhanced significantly by the presence of 0.02–0.5 ng/ml teleocidin (*H. Yamamoto* et al. 1981b; *Hoshino* et al. 1981). These data further support the theory that teleocidin and TPA exert their effects by the same mechanism(s). Recently discovered tumor promoters like lyngbyatoxin A, aplysiatoxin, and palytoxin, already shown to be active inducers of EBV, like TPA, may act in the same way in the transformation of human lymphocytes by EBV. Preliminary results showed that activated serum factor was also able to enhance transformation in the EBV system (*Bauer* et al., unpublished observation). It is, therefore, essential to test whether this physiological factor shares the same properties with TPA in vivo.

Henderson et al. (1975) and *Henderson* and *Ribecky* (1980) reported that the frequency of spontaneous transformation of explanted leukocytes obtained from EBV-seropositive adults can be increased by treatment of the cells in vitro with the carcinogen MNNG. Autologous EBV-antibody-containing serum inhibited transformation after MNNG treatment, as did TPA. This again showed that at least cell-free virus is involved in this transformation process. However, they showed that MNNG enhancement of transformation was also detected in the soft agarose colony-forming assay (*Yamamoto* and *Hinuma* 1976). Moreover, the effect was greatest in the first 24 h of virus exposure. These data indicate that MNNG can influence some early event in the EBV-leukocyte interaction, before EBV-induced host cell DNA replication begins. These results are consistent with Casto's postulation (*Casto* 1973; *Casto* et al. 1976, 1979) that an increase in viral transformation might be expected if viral genetic material were available or replicated during the time when lesions are present in the cell DNA. However, no enhancement of viral transformation was detected following treatment with other chemical or physical mutagens, such as methanesulfonate, UV light, or X-radiation (*Henderson* and *Ribecky* 1980).

4.2 Other Viral Systems

Fisher et al. extensively studied the modulation of adenovirus transformation and replication by tumor promoters (*Fisher* et al. 1978, 1979b, 1981). One of the greatest advantages of this system is that one can compare the effects of tumor promoters with those of other initiating agents, such as chemical or physical carcinogens. TPA (100 ng/ml) caused a two- to threefold enhancement of transformation of rat embryo (RE) cells by a temperature-sensitive mutant of Ad5 (H5 ts 125). Analogous to the EBV system, transformed foci appeared earlier and were larger in the presence of TPA than were controls. The addition of TPA could be delayed for up to 1 week, suggesting that this chemical enhances transformation by facilitating the expression of the transformed state, rather than by its effect on virus uptake or integration, which might occur within 72 h. Enhanced transformation of these cells by H5 ts 125 virus was also seen after exposure to the carcinogens 7-, 12,-dimethylbenzanthracene or benzopyrene. This enhancement was further augmented two- to threefold when TPA was included.

Enhancement by TPA of transformation of rodent cells by Ad5 is, to some extent, similar to that with X-radiation, since it was possible to add TPA after the treatment with carcinogenic agents in both systems. This situation represents a clear difference from that of carcinogens in which chemical treatment prior to or during viral infection is a prerequisite for the synergy (*Casto* 1973; *Diamond* et al. 1974; *Hirai* et al. 1974; *Casto* et al. 1976; *Fisher* 1983; *Milo* et al. 1978).

A similar increase in transformed foci was obtained with 2 μg/ml melittin, a bee venom polypeptide, or 10 ng/ml EGF (*Fisher* et al. 1978, 1979a, 1981). These chemicals possess a number of the biological features of TPA, including the tumor-promoting activity in mouse skin and appear to act initially on the cell membrane, as does TPA (*Rose* et al. 1976; *Lee* and *Weinstein* 1978b). These findings suggest that alterations in cell membrane functions may play an important role in the enhancement of transformation by adenovirus. These membrane changes may lead to the stimulation of viral RNA and viral replication, which have been shown to be associated with the enhancement of mouse cell transformation by TPA.

Anchorage independence, namely, the ability to grow in semisolid medium in vitro has been found to correlate best with in vivo tumorigenicity, with some exceptions (*Fisher* 1983). TPA has also been shown to enhance H5 ts 125-transformed cells on agar (*Fisher* et al. 1979a,b). This is not restricted to the adenovirus system, since TPA has been shown to stimulate anchorage independence in carcinogen-treated mouse epidermal cells. With a series of diterpens, there is a good correlation between tumor-promoting activity on mouse skin and enhanced anchorage independence of RE clones. Interestingly, the acquisition of anchorage independence seems to be irreversible, unlike the majority of TPA effects (*Fisher* 1983). In this system, transformed cells do not spontaneously grow in agar in the early passages, but gain the ability to do so with an increase in time. However, nucleic acid hybridization analysis of the various clones obtained after both early and late passages revealed that alterations in the abundance or state of integration of adenovirus sequences did not appear to be involved in the mechanism for the acquirement of anchorage independence.

TPA increased the frequency of transformation of mouse 3T3 cells by polyoma virus 8- to 40-fold at a very high concentration of the drug (500 ng/ml). The same enhancement was seen in cells treated with griseofulvin, melittin, EGF, vinblastin, cytochalasin B, podophyllotoxin, colcemid and colchicin (*Seif* 1980). Common to all these chemicals is their ability to disorganize microtubules or microfilaments. Since both small t antigen and griseofulvin were reported to disorganize the actin microfilament/microtubules systems, the author speculated that small t may act as a promoter and that large T may act as an initiator.

In the Ad5 transformation system, the continuous presence of TPA is required to achieve the maximum effect (*Fisher* 1983). However, TPA effect on polyoma viral transformation was observed only when it was applied for the first 7 days after infection. It is, therefore, conceivable that TPA acts on the early stage, rather than affecting a later stage, of polyoma virus transformation.

There is also a report on the stimulation of retroviral transformation by TPA (*Miyaki* et al. 1980). When TPA was added at 1 ng/ml to the culture medium 3 days after infection of normal human fibroblast cells with MSV(MLV), the transformed cells in the TPA-treated cultures were rounder and the foci were more refractory than those in untreated cultures. A greater than twofold increase in foci was observed in treated cultures compared with untreated cultures.

5 Screening of Tumor Promoters with EBV System

A number of short-term assay systems for initiating agents have been reported. As an example, the Ames test has been widely used (*Ames* et al. 1973). In these test systems, DNA damage, mutagenicity, and chromosomal abnormalities were used as indicators of the activities of initiating agents. However, similar systems for tumor promoters have not yet been completed. It is certainly necessary to develop such assay systems for tumor promoters, in addition to those for initiating agents, since most human cancers are believed to arise from multifactorial interactions, as already described. For this purpose, *Fujiki* et al. (1983) have tried to develop a short-term screening assay system for tumor promoters using various biological activities, such as: 1) irritation on mouse ears; 2) induction of ODC; 3) aggregation induction and differentiation of HL-60 cells, a promyelocytic leukemia cell line of human origin; 4) aggregation of EBV-transformed lymphoblastoid cell lines; 5) inhibition of differentiation of Friend erythroleukemia cells.

Since the EBV-inducing activity of tumor promoters correlates, to some extent, with the tumor-promoting activity of certain drugs (*zur Hausen* et al. 1979), this system may provide a useful tool for a convenient and short-term in vitro assay for certain types of tumor promoter. Indeed, *Ito* et al. (1981) developed such short-term assays based on the EBV superinduction system, which were composed of nonproducer Raji cells as the indicator cells, *n*-butyrate as the EBV inducer, and the test substance. These studies revealed that EBV activators were widely distributed among the plants of the euphorbiaceae and thymelaeceae families. These activators were diterpene esters based on tigliane, ingenane, and daphnane hydrocarbon skeltons. They were effective in inducing EA in Raji cells at the nonogram per milliliter level. Sixteen of 500 such plants were shown to contain active substances for EBV induction. Moreover, they also enhanced the transformation of normal human leukocytes by EBV (*Mizuno* et al. 1983) in a fashion similar to TPA (*Yamamoto* et al. 1979). Interestingly, in the areas where two EBV-associted malignancies, BL and NPC, are endemic, people use many plants with these active substances as common folk remedies.

In addition to deterpene esters, various chemicals have tumor-promoting activity in mouse skin – fatty acids, phenols, nonaromatic hydrocarbons, cigarette smoke condensate, etc. Another class of chemicals known to promote tumor formation in tissues other than skin includes phenols, TPA, fatty acids, phenobarbital, saccharin, cyclamate, and bile acids (*Hozumi* 1978). It is important to test in this system whether these promoters, especially naturally occurring ones, from exogenous and endogenous sources, such as bile acids or growth hormones, cause EBV-inducing activity.

6 Concluding Remarks

As is well known, the BL cell line is characterized by aneuploidy with a chromosome 14 marker, a potential for tumor formation in nude mice, and derivation from a single-cell clone. On the other hand, cell lines derived spontaneously from normal, or by in vitro immortalization of human B lymphocytes by EBV, are of polyclonal derivation, diploid, and have no potential to form tumors in nude mice (*Nilsson* 1979). These facts suggest that cell lines immortalized in vitro by EBV are not true malignant cells and require additional genetic alteration to become malignant. Furthermore, it has long been suspected that the reason why BL and NPC are endemic for black children in Africa and for adults in South China, respectively, as follows: Although EBV may be a prerequisite for or facilitate the development of malignancy, one or more other factors (mechanisms) are needed. One can assume such factors to be environmental (exogenous) and endogenous (e.g., natural resistance or susceptibility to microbial infection). It is most likely that viral carcinogenesis, like chemical carcinogenesis, results from the interaction of a virus with various chemical, physical, and biological factors.

In this context, I have described the in vitro studies on the interaction of viruses with tumor promoters with special reference to the EBV system. A potent tumor promoter, TPA, caused efficient induction of EBV in persistently infected lymphoblastoid cell lines and enhancement of transformation of normal human leukocytes by EBV. A very important question is whether these events are related. As already described, TPA exerts its primary effect at the cell surface and is considered to affect nuclear functions by transmembrane signals, as mediated by the phosphorylation of specific cellular proteins. Although it has yet to be determined whether this model is also applicable to EBV induction by TPA, the chemical exerts its effect by enhancing the transcription of EBV genomes, which results in increased production of EBV-specific antigens and virus particles. Regarding the transformation of human leukocytes by EBV, it has clearly been shown that enhancement by TPA results from three independent mechanisms: First, a direct effect on the early stage of transformation; second, the induction of biologically active EBV and consequent infection of coexisting target B cells; and third, abrogation of T-cell activity, which suppresses EBV-induced transformation. The first and second of these mechanisms indicate that the enhancement of transformation and the induction of persisting viral genomes are related, since these events occur through alteration of the program of viral and/or host gene expression. The third mechanism is more important in vivo than in vitro. Although infection with EBV alone is not sufficient to cause malignancy under normal conditions, it is likely that various environmental factors combine to generate lymphomas by inducing infectious virus on one hand and suppressing immunological resistance of the host on the other. Induction of infectious virus increases the pool size of EBV-infected cells. These cells might play a role as "conditioned cells" (*zur Hausen* 1980), which could facilitate occurrence of subsequent translocation in the 14 chromosome by other factors, leading to a monoclonal expansion of genetically altered cells.

Acknowledgments. I would like to thank Harald zur Hausen for his invaluable discussion and critical reading, James R. Blakeslee for the English, and Kazunori Hirayoshi for the preparation of this review. I also thank P.B. Fisher for providing his reviews in preprint form. Our studies described in this manuscript were performed in fruitful collaboration with H. zur Hausen, G. Bauer, N. Mueller-Lantzsch, and K. Bister, in the Institut für Virologie, Zentrum für Hygiene, University of Freiburg, and Y. Hinuma in the Institute for Virus Research, Kyoto University.

Various aspects of the studies described in this review were supported by the Alexander von Humboldt Stiftung, by the Deutsche Forschungsgemeinschaft and by Grants-in-Aid for Cancer Research from the Ministry of Education, Science, and Culture, and the Ministry of Health and Welfare, Japan.

References

Adams A, Lindahl T, Klein G (1973) Linear association between cellular DNA and Epstein-Barr virus DNA in a human lymphoblastoid cell line. Proc Natl Acad Sci USA 70:2888–2892

Ames BN, Durston WE, Yamasaki E, Lee FD (1973) Carcinogens are mutagens: A simple test system combining liver homogenates for activation and bacteria for detection. Proc Natl Acad Sci USA 70:2281–2285

Amtmann E, Sauer G (1982) Activation of non-expressed bovine papilloma virus genomes by tumor promoters. Nature 296:675–677

Anisimova E, Saemundsen AK, Roubal J, Vonka V, Klein G (1982) Effects of n-butyrate on Epstein-Barr virus-carrying lymphoma lines. J Gen Virol 58:463–471

Arya SK (1980) Phorbol ester-mediated stimulation of the synthesis of mouse mammary tumor virus. Nature 284:71–72

Ashendel CL, Staller JM, Boutwell RK (1983) Identification of a calcium- and phospholipid-dependent phorbol ester binding activity in the soluble fraction of mouse tissues. Biochem Biophys Res Commun 111:340–345

Bauer G (1983a) Quantitative analysis of the cooperative effect between inducers of Epstein-Barr virus antigen synthesis. J Gen Virol 64:1337–1346

Bauer G (1983b) Induction of Epstein-Barr virus early antigens by corticosteroids: Inhibition by TPA and retinoic acid. Int J Cancer 31:291–295

Bauer G, Hofler P, Simon M (1982a) Epstein-Barr virus induction by a serum factor. II. Purification of a high molecular weight protein that is responsible for induction. J Biol Chem 257:11405–11410

Bauer G, Hofler P, Simon M (1982b) Epstein-Barr virus induction by a serum factor. III. Characterization of the purified molecule and the mechanism of its activation. J Biol Chem 257:11411–11415

Bauer G, Hofler P, zur Hausen H (1982c) Epstein-Barr virus induction by a serum factor. I. Induction and cooperation with additional inducers. Virology 121:184–194

Bayliss GJ, Nonoyama M (1978) Mechanism of infection with Epstein-Barr virus. III. The synthesis of protein in superinfected Raji cells. Virology 87:204–207

Ben-Sasson SA, Klein G (1981) Activation of the Epstein-Barr virus genome by 5-azacytidine in latently infected human lymphoid lines. Int J Cancer 28:131–135

Berenblum I (1954) A speculative review: the probable nature of promoting action and its significance in the understanding of the mechanism of carcinogenesis. Cancer Res 14:471–477

Berenblum I (1975) Origin of the concept of sequential stages of skin carcinogenesis. In: Becker FF (ed) Cancer 1. Plenum, New York, pp 323:344

Bishop JM, Varmus H (1982) Function and origins of retroviral transforming genes. In: Weiss R, Teich N, Varmus H, Coffin J (eds) RNA tumor viruses. Cold Spring Harbor Laboratory, New York, pp 999–1108

Birnboim HC (1982) DNA strand breakage in human leukocytes exposed to a tumor promoter, phorbol myristate acetate. Science 215:1247–1250

Bister K, Yamamoto N, zur Hausen H (1979) Differential inducibility of Epstein-Barr virus in cloned, non-producer Raji cells. Int J Cancer 23:818–825

Boutwell RK (1974) The function and mechanism of promoters of carcinogenesis. CRC Crit Rev Toxicol 2:419–443

Brown KD, Dicker P, Rozengurt E (1979) Inhibition of epidermal growth factor binding to surface receptors by tumor promoters. Biochem Biophys Res Commun 86: 1037–1043

Brugge JS, Erikson RL (1977) Identification of a transformation-specific antigen induced by an avian sarcoma virus. Nature 269:346–348

Brugge J, Erikson E, Collett MS, Erikson RL (1978) Peptide analysis of the transformation-specific antigen from avian sarcoma virus-transformed cells. J Viol 26:773–782

Castagna M, Takai Y, Kaibuchi K, Sano K, Kikkawa U, Nishizuka Y (1982) Direct activation of calcium-activated, phospholipid-dependent protein kinase by tumor-promoting phorbol esters. J Biol Chem 257:7847–7851

Casto BC (1973) Enhancement of adenovirus transformation by treatment of hamster cells with ultraviolet-irradiation, DNA base analogs, and dibenz(a,h)-anthracene. Cancer Res 33:402–407

Casto BC, Pieczynski WJ, Janosko N, DiPaolo JA (1976) Significance of treatment interval and DNA repair in the enhancement of viral transformation by chemical carcinogens and mutagens. Chem Biol Interact 13:105–125

Casto BC, Miyagi M, Meyers J, DiPaolo JA (1979) Increased integration of viral genome following chemical and viral treatment of hamster embryo cells. Chem Biol Interact 25:255–269

Chang RS, Hsieh MW, Blankenship W (1971) Initiation and establishment of lymphoid cell lines from blood of healthy persons. J Natl Cancer Inst 47:469–477

Cheng YV, Chen JY, Glaser R, Henle W (1980) Frequency and levels of antibodies to Epstein-Barr virus specific DNase are elevated in patients with nasopharyngeal carcinoma. Proc Natl Acad Sci USA 77:6162–6165

Clough W (1979) Deoxyribonuclease activity found in Epstein-Barr virus producing lymphoblastoid cells. Biochemistry 18:4517–4521

Clough W (1980) An endonuclease isolated from Epstein-Barr virus-producing human lymphoblastoid cells. Proc Natl Acad Sci USA 77:6194–6198

Clough W, McMahon J (1981) Characterization of the Epstein-Barr virion-associated DNA polymerase as isolated from superinfected and drug-stimulated cells. Biochim Biophys Acta 656:76–85

Colburn NH, Vorder Bruegge WF, Bates J, Yuspa SH (1978) Epidermal cell transformation in vitro, in mechanism of tumor promotion and cocarcinogenesis. In: Slaga TJ, Sivak A, Boutwell RK (eds) Carcinogenesis, vol 2. Raven, New York, pp 257–271

Collett MS, Erikson RL (1978) Protein kinase activity assocaited with the avian sarcoma virus src gene product. Proc Natl Acad Sci USA 75:2021–2024

Collett MS, Brugge JS, Erikson RL (1978) Characterization of a normal avian cell protein related to the avian sarcoma virus transforming gene product. Cell 15:1363–1369

Colletta G, DiFiore PP, Ferrentino M, Pietropaolo C, Turco MC, Vecchio G (1980) Enhancement of viral gene expression in Friend erythroleukemic cells by 12-O-tetradecanoyl-phorbol-13-acetate. Cancer Res 40:3369–3373

Cushing RT, Morgan HR (1952) Effect of some metabolic analogs on the growth of mumps and influenza viruses in tissue cultures. Proc Soc Exp Biol 79:497–500

Datta AK, Colby JE, Shaw JE, Pagano JS (1980a) Acyclovir inhibition of Epstein-Barr virus replication. Proc Natl Acad Sci USA 77:5163–6166

Datta AK, Feighny RJ, Pagano JS (1980b) Induction of Epstein-Barr virus-associated DNA polymerase by 12-O-tetradecanoylphorbol-13-acetate. Purification and characterization. J Biol Chem 255:5120–5125

Diala ES, Hoffman RM (1983) Epstein-Barr Hr-1 virion DNA is very highly methylated. J Virol 45:482–483

Diamond L, Knorr R, Shimizu Y (1974a) Enhancement of simian virus 40-induced transformation of Chinese hamster embryo cells by 4-nitroquinoline-1-oxide. Cancer Res 34:2599–2604

Diamond L, O'Brien S, Donaldson C, Shimizu Y (1974b) Growth stimulation of human diploid fibroblasts by the tumor promoter 12-O-tetradecanoyl-phorbol-13-acetate. Int J Cancer 13:721

Diamond L, O'Brien TG, Rovera G (1977) Inhibition of adipose conversion of 3T3 fibroblasts by tumor promoters. Nature 269:247–249

Diamond L, O'Brien TG, Rovera G (1978) Tumor promoters: effects on proliferation and differentiation in cell culture. Life Sci 23:1979–1988

Diamond L, O'Brien TG, Baird WM (1980) Tumor promoters and the mechanism of tumor promotion. Adv Cancer Res 32:1–74

Driedger PE, Blumberg PM (1977) The effect of phorbol diesters on chicken embryo fibroblasts. Cancer Res 37:3257–3265

Diedger PE, Blumberg PM (1980) Specific binding of phorbol ester tumor promoters. Proc Natl Acad Sci USA 77:567

Ehrlich M, Wang RH (1981) 5-methylcytosine in eukaryotic DNA. Science 212:1350–1357

Eliasson L, Kallin B, Patarroyo M, Klein G, Fujiki H, Sugimura T (1983a) Activation of the EBV-cycle and aggregation of human blood lymphocytes by the tumor promoters teleocidin, lyngbyatoxin A, aplysiatoxin and debromoaplysiatoxin. Int J Cancer 31:7–11

Eliasson L, Nilsson M, Klein G (1983b) Tumor promoter induced demethylation in the EBV-genome, abstract. 14th Int symp of The Princess Takamatsu Cancer Res Fund on cellular interactions by environmental tumor promoters, pp 26–27

Epstein M, Achong BG (1977) Recent progress in Epstein-Barr virus research. Annu Rev Microbiol 31:421–445

Erikson E, Collett MS, Erikson E, Purchio AF (1979) Evidence that the avian sarcoma virus transforming gene product is a cyclic AMP-independent protein kinase. Proc Natl Acad Sci USA 76:6260–6264

Ernberg I, Moar MH (1981) Induction of Epstein-Barr virus early antigen, virus capsid antigen and virus DNA synthesis by mitomycin C. J Gen Virol 52:39–48

Ernberg I, Andersson-Anvret M, Klein G, Lundin L, Killander D (1977) Relationship between amount of Epstein-Barr virus-determined nuclear antigen per cell and number of EBV-DNA copies per cell. Nature 266:269–270

Feighny RJ, Henry BE, Pagano JS (1981) Epstein-Barr virus-induced deoxyribonuclease and the utilization of host-cell DNA degradation products in viral DNA replication. Virology 115:395–400

Fisher PB (1983) Enhancement of viral transformation and expression of the transformed phenotype by tumor promoters. In: Slaga (ed) Mechanisms of tumor promotion: Tumor promotion and cocarcinogenesis in vitro. CRC, Boca Raton
Fisher PB, Weinstein IB, Eisenberg D, Ginsberg HS (1978) Initiations between adenovirus, a tumor promoter, and chemical carcinogens in transformation of rat embryo cell cultures. Proc Natl Acad Sci USA 75:2311–2314
Fisher PB, Bozzone JH, Weinstein IB (1979a) Tumor promoters and epidermal growth factor stimulate anchorage-independent growth of adenovirus transformed rat embryo cells. Cell 18:695
Fisher PB, Dorsch-Hasler K, Weinstein IB, Ginsberg HS (1979b) Tumor promoters enhance anchorage-independent growth of adenovirus-transformed cells without altering the integration pattern of viral sequences. Nature 281:591
Fisher PB, Young CSH, Weinstein IE, Carter TH (1981) Tumor promoters alter the temporal program of adenovirus replication in human cells. Mol Cell Biol 1:370
Fisher PB, Weinstein IB, Goldstein NI, Young CSH, Carter TH (1982) Modulation of adenovirus transformation and replication by chemical carcinogens and phorbol ester tumor promoters. In: Rich MA, Furmanski P (eds) Biological carcinogenesis. Dekker, New York, pp 261–290
Freeman AE, Price PJ, Igel HJ, Young JC, Maryak JM, Huebner RJ (1970) Morphological transformation of rat embryo cells induced by diethylnitrosamine and murine leukemia viruses. J Natl Cancer Inst 44:65–78
Fresen KO, Cho MS, zur Hausen H (1978) Recovery of transforming EBV from nonproducer cells after superinfection with non-transforming P3HR-1 EBV. Int J Cancer 22:378–383
Fujiki H (1981) Teleocidin – New tumor promoter (in Japanese). Taisha 18:785–792
Fujiki H, Mori M, Nakayasu M, Terada M, Sugimura T (1979) A possible naturally occurring tumor promoter, teleocidin B from Streptomyces. Biochem Biophys Res Comm 90:976–983
Fujiki H, Suganuma M, Tahira T, Yoshioka A, Nakayasu M, Hayashi K, Endo Y, Shudo K, Takayama S, Moore RE, Sugimura T (1983) New classes of tumor promoter: teleocidin, aplysiatoxin and palytoxin, abstract. 13th Int symp of The Princess Takamatsu Cancer Res Fund on cellular interactions by environmental tumor promoters, pp 58–59
Fusenig NE, Samsel W (1978) Mechanisms of tumor promotion and cocarcinogenesis. In: Slaga TJ, Sivak A, Boutwell RK (eds) Carcinogenesis, vol II. Raven, New York, pp 203–220
Gerber P (1972) Activation of Epstein-Barr virus by 5-bromo-deoxyuridine in virus free human cells. Proc Natl Acad Sci USA 69:83–85
Gerber P, Whang-Peng J, Monroe JH (1969) Transformation and chromosome changes induced by Epstein-Barr virus in normal human leukocyte cultures. Proc Natl Acad Sci USA 63:740–747
Goldberg AR, Delclos KB, Blumberg PM (1980) Phorbol ester action is independent of viral and cellular src kinase levels. Science 208:191–193
Gottesman MM, Sobel ME (1980) Tumor promoters and Kirsten sarcoma virus increase synthesis of a secreted glycoprotein by regulating levels of translatable mRNA. Cell 19:449–455
Grossberger D, Clough W (1981) Characterization of purified Epstein-Barr virus induced deoxyribonucleic acid polymerase: Nucleotide turnover, processiveness, and phosphonoacetic acid sensitivity. Biochemistry 20:4049–4055
Hamburger WW, White CP, Brown RW (1981) Effect of epidermal growth factor on proliferation of human tumor cells in soft agar. J Natl Cancer Inst 67:825–830
Hampar B, Derge JG, Martos LM, Walker JL (1972) Synthesis of Epstein-Barr virus after activation of the viral genome in a virus negative human lymphoblastoid cell (Raji) made resistant to 5-bromodeoxyuridine. Proc Natl Acad Sci USA 69:48–82
Hanafusa H (1977) Cell transformation by RNA tumor viruses. In: Fraenkel-Conrat H, Wagner RR (eds) Comprehensive virology, vol 10. Plenum, New York, pp 401–483

Harada S, Katsuki T, Yamamoto H, Hinuma Y (1981) Mechanism of enhancement of Epstein-Barr virus induced transformation of peripheral blood lymphocytes by a tumor promoter, TPA, with special reference to lowering of cytotoxicity of T cells. Int J Cancer 27:617–623

Harrison J, Auersperg N (1981) Epidermal growth factor enhances viral transformation of granulosa cells. Science 213:218–219

Hecker E (1968) Cocarcinogenic principles from the seed oil of Croton tiglium and from other euphorbiaceae. Cancer Res 28:2338–2349

Hecker E (1975) Cocarcinogens and cocarcinogenesis. In: Grundman E (ed) Handbuch der allgemeinen Pathologie, vol 6/6, Geschwulste/Tumors II. Springer, Berlin Heidelberg New York, pp 651–676

Hecker E (1978) Mechanisms of tumor promotion and cocarcinogenesis. In: Slaga TJ, Sivak A, Boutwell RK (eds) Carcinogenesis, vol II. Raven, New York, pp 11–48

Hellman KB, Hellman A (1981) Induction of type-C retrovirus by the tumor promoter TPA. Int J Cancer 27:95–100

Henderson EE, Ribecky R (1980) Transformation of human leukocytes with Epstein-Barr virus after cellular exposure to chemical or physical mutagens. J Natl Cancer Inst 64:33–39

Henderson EE, Norin AJ, Strauss BS (1975) Induction of permanently proliferating human lymphoblastoid lines by N-methyl-N'-nitro-N-nitrosoguanidine. Cancer Res 35:358–363

Henle G, Henle W, Diehl V (1968) Relation of Burkitt tumor associated herpes-type virus to infectious mononucleosis. Proc Natl Acad Sci USA 59:94–101

Henle W, Henle G (1974) Epstein-Barr virus: The cause of infectious mononucleosis – a review. In: Biggs PM, de-The G, Payne LN (eds) Oncogenesis and herpes-viruses. IARC, Lyon, 269–274

Henle W, Diehl V, Kohn G, zur Hausen H, Henle G (1967) Herpes-type virus and chromosome marker in normal leucocytes after growth with irradiated Burkitt cells. Science 157:1064–1065

Henle W, Henle G, Zajac BA, Pearson GM, Waubke R, Scriba M (1970) Differential reactivity of human serums with early antigens induced by Epstein-Barr virus. Science 169:188–190

Hiatt HH, Watson JD, Winsten JA (1977) Origin of human cancer. Cold Spring Harbor conferences on cell proliferation, vol 4, A, B, C

Hinuma Y, Konn M, Yamaguchi J, Wudarski DJ, Blakeslee JR, Grace JT (1967) Immunofluorescence and herpes-type virus particles in the P3HR-1 Burkitt lymphoma cell line. J Virol 1:1045–1051

Hinuma Y, Nagata K, Hanaoka M, Nakai M, Matsumoto T, Kinoshita K, Shirakawa S, Miyoshi I (1981) Adult T-cell leukemia: antigen in an ATL cell line and detection of antibodies to the antigen in human sera. Proc Natl Acad Sci USA 78:6476–6480

Hirai K, Defendi V, Diamond L (1974) Enhancement of simian virus 40 transformation and integration by 4-nitroquinoline-1-oxide. Cancer Res 34:3497–3500

Hoffman-Liebermann B, Liebermann D, Sachs L (1981) Regulation of gene expression by tumor promoters. III. Complementation of the developmental program in myeloid leukemic cells by regulating mRNA production and mRNA translation. Int J Cancer 28:615–620

Hollenberg MD, Cuatrecasas P (1973) Epidermal growth factor: receptors in human fibroblasts and modulation of action by cholera toxin. Proc Natl Acad Sci USA 70: 2964–2968

Hoshino H, Miwa M, Fujiki H, Sugimura T (1980) Aggregation of human lymphoblastoid cells by tumor-promoting phorbol esters and dihydroteleocidin B. Biochem Biophys Res Commun 95:842–848

Hoshino H, Miwa M, Fujiki H, Sugimura T, Yamamoto H, Katsuki T, Hinuma Y (1981) Enhancement of Epstein-Barr virus-induced transformation of human lymphocytes by teleocidin. Cancer Lett 13:275–280

Hoshino H, Miwa M, Fujiki H (1983) A new tumor promoter, dihydroteleocidin B, enhances cell growth and the production of murine leukemia virus by fibroblasts. Int J Cancer 31:509–513
Hozumi M (1978) On anticarcinogens and cocarcinogens (in Japanese). Tanpakushitsu Kakusan Koso 23:460–468
Hudewentz J, Bornkamm GW, zur Hausen H (1980) Effect of the diterpene ester TPA on Epstein-Barr virus antigen and DNA synthesis in producer and nonproducer cell lines. Virology 100:175–178
Hummel M, Kieff E (1982) Epstein-Barr virus RNA. 8. Viral RNA in permissively infected B95-8 cells. J Virol 43:262–272
Ishii D, Fibach E, Yamasaki H, Weinstein IB (1978) Tumor promoters inhibit morphological differentiation in cultured mouse neuroblastoma cells. Science 200: 556–559
Ito Y, Yanase S, Fijiki J, Harayama T, Takashima M, Imanaka H (1981) A short term in vitro assay for promoter substances using human lymphoblastoid cells latently infected with Epstein-Barr virus. Cancer Lett 13:29–37
Jones PA, Taylor SM (1980) Cellular differentiation, cytidine analogs and DNA methylation. Cell 20:85–93
Kallin B, Klein G (1983) Epstein-Barr virus carried by Raji cells: a mutant in early functions? Intervirology 19:47–51
Kallin B, Luka J, Klein G (1979) Immunochemical characterization of Epstein-Barr virus-associated early and late antigens in n-butyrate-treated P3HR-1 cells. J Virol 32:710–716
Kaschka-Dierich C, Adams A, Lindahl T, Bornkamm GW, Bjursell G, Klein G, Giovanella BC, Singh S (1976) Intracellular forms of Epstein-Barr virus DNA in human tumor cells in vivo. Nature 260:302–306
Katsuki T, Hinuma Y (1976) A quantitative analysis of the susceptibility of human leukocytes to transformation by Epstein-Barr virus. Int J Cancer 18:7–13
Katsuki T, Hinuma Y, Yamamoto N, Abo T, Kumagai K (1977) Identification of the target cells in human B lymphocytes for transformation by Epstein-Barr virus. Virology 83:287–294
Kawanishi M, Ito Y (1983) Induction of EB virus-specific DNase activity by the treatment of nonproducer Raji cells with chemical inducers (in Japanese), abstract. 31st Meeting of The Society of Jap. Virologists, p 1031
Kawanishi M, Sugawara K, Ito Y (1981) Epstein-Barr virus polypeptides: a comparative study with superinfected Raji, IUdR-treated and n-butyrate-treated P3HR-1 cells. Virology 109:72–81
Keir HM, Gold E (1963) Deoxyribonucleic acid nucleotidyltransferase and deoxyribonuclease from cultured cells infected with herpes simplex virus. Biochim Biophys Acta 72:263–276
Kennedy A, Mondal S, Heidelberger C, Little JB (1978) Enhancement of X-radiation transformation by a phorbol ester using C3H/10T/1/2 CI 8 mouse embryo fibroblasts. Cancer Res 38:2896–2899
Kintner C, Sugden B (1981) Conservation and progressive methylation of Epstein-Barr viral DNA sequences in transformed cells. J Virol 38:305–316
Klein G (1973) The Epstein-Barr virus. In: Kaplan AS (ed) The herpesviruses. Academic, New York, pp 521–555
Klein G, Dombos L (1973) Relationship between sensitivity of EBV-carrying lymphoblastoid lines to superinfection and the inducibility of the resident viral genome. Int J Cancer 11:327–337
Klein G, Zeuthen J, Eriksson I, Terasaki P, Bernoco M, Rosen A, Masucci G, Povey S, Ber R (1980) Hybridization of a myeloid leukemia derived cell line (K562) with a Burkitt lymphoma line (P3HR-1). J Natl Cancer Inst 64:725–735
Krieg P, Amtmann E, Sauer G (1981) Interaction between latent papovavirus genomes and the tumor promoter TPA. FEBS Lett 128:191–194

Larocca D, Clough W (1982) Hypomethlyation of Epstein-Barr virus DNA in the non-producer B-cell line EBR. J Virol 43:1129–1131
Lasne C, Gentil A, Chouroulinkov I (1974) Two-stage malignant transformation of rat fibroblasts in tissue culture. Nature 247:490–491
Lau RY, Nonoyama M, Klein G (1981) Somatic cell hybrids between human lymphoma and human myeloid leukemia cells. 1. Induction of resident viral genome transcription and viral antigen formation by IUdR and TPA. Virology 110:259–269
Lee LS, Weinstein IB (1978a) Epidermal growth factor, like phorbol esters, induces plasminogen activator in HeLa cells. Nature 274:696–597
Lee LS, Weinstein IB (1978b) Tumor-promoting phorbol esters inhibit binding of epidermal growth factor to cellular receptors. Science 202:313–315
Lee LS, Weinstein IB (1978c) Uptake of the tumor promoting agent 12-O-tetradecanoyl-phorbol-acetate by HeLa cells. J Environ Pathol Toxicol 1:627–639
Leibold W, Flanagan TD, Menezes J, Klein G (1975) Induction of Epstein-Barr virus-associated nuclear antigen during in vitro transformation of human lymphoid cells. J Natl Cancer Inst 54:65–68
Lenoir G (1979) Unit of biological carcinogens: laboratory studies. In: International agency for research on cancer annual report 1979. WHO/IARC, Lyon, pp 83–84
Levinson AD, Oppermann H, Levintow L, Varmus HE, Bishop JM (1978) Evidence that the transforming gene of avian sarcoma virus encodes a protein kinase associated with a phosphoprotein. Cell 15:561–572
Lin JC, Shaw JE, Smith MC, Pagano JS (1979) Effect of 12-O-tetradecanoyl-phorbol-13-acetate on the replication of Epstein-Barr virus. 1. characterization of viral DNA. Virology 99:183–187
Lin JC, Smith MC, Pagano JS (1981) Induction of replication of Epstein-Barr virus DNA by 12-O-tetradecanoyl-phorbol-13-acetate. II. Inhibition by retinoic acid and 9-(2-hydroxyethoxyametyl)-guanine. Virology 111:294–298
Lin JC, Smith MC, Pagano JS (1983a) Activation of latent Epstein-Barr virus genomes: selective stimulation of synthesis of chromosomal proteins by a tumor promoter. J Virol 45:985–991
Lin JC, Smith MC, Pagano JS (1983b) Enhancement of replication of Epstein-Barr virus DNA in human lymphoblastoid cell lines by N-methyl-N'-nitro-N-nitrosoguanidine. Carcinogenesis 4:185–188
Lindahl T, Adams A, Bjursell B, Bornkamm GW, Kaschka-Dierich C, Jehn U (1976) Covalently closed circular duplex DNA of Epstein-Barr virus in a human lymphoid cell line. J Mol Biol 102:511–530
Lipp M, Scherer B, Lips G, Brandner G, Hunsmann G (1982) Diverse effects: augmentation, inhibition, and non-efficacy of 12-O-tetradecanoylphorbol-13-acetate (TPA) on retrovirus genome expression in vivo and in vitro. Carcinogenesis 3:261–265
Lotan R (1980) Effects of vitamin A and its analogs (retinoids) on normal and neoplastic cells. Biochim Biophys Acta 605:33–91
Luka J, Kallin B, Klein G (1979) Induction of the Epstein-Barr virus (EBV) cycle in latently infected cells by n-butyrate. Virology 94:228–231
MacPherson I (1973) In: Tissue cultures. Academic, New York
Magrath IT, Pizzo PA, Novikovs L, Levine A (1979) Enhancement of Epstein-Barr virus replication in producer cell lines by a combination of low temperature and coricosteroids. Virology 97:477–481
Meyn MS, Rossman T, Troll W (1977) A protease inhibitor blocks SOS functions in Escherichia coli: antipain prevents repressor inactivation, ultra-violet mutagenesis and filamentous growth. Proc Natl Acad Sci USA 74:1152–1156
Miao RM, Fieldsteel AH, Fodge DW (1978) Opposing effects of tumor promoters on erythroid differentiation. Nature 274:271–272
Miller G, Lipman M (1973) Release of infectious Epstein-Barr virus by transformed marmoset leukocytes. Proc Natl Acad Sci USA 70:190–194

Miller G, Lisco H, Kohn HI, Stitt D (1971) Establishment of cell lines from normal adult human blood leukocytes by exposure to Epstein-Barr virus and neutralization by human sera with Epstein-Barr virus antibody. Proc Soc Exp Biol Med 137: 1459–1465

Miller RL, Rapp F (1977) Studies of an Epstein-Barr virus-induced DNA polymerase. Virology 76:494–502

Milo GE, Blakeslee JR, Hart R, Yohn DS (1978) Chemical carcinogen alteration of SV40 virus induced transformation of normal human cell populations in vitro. Chem Biol Interact 22:185–197

Miyaki M, Akamatsu N, Rokutanda M, Oono T, Yoshikura H, Sasaki M, Tonomura A, Utsunomiya J (1980) Increased sensitivity of fibroblasts of skin from patients with adenomatosis coli and Peutz-Jeghers' syndrome to transformation by murine sarcoma virus. Gann 71:797–803

Mizuno F, Aya T, Osato T (1976) Growth in semisolid agar medium of human cord leukocytes freshly transformed by Epstein-Barr virus. J Natl Cancer Inst 56:171–173

Mizuno F, Koizumi S, Osato T, Kokwaro JO, Ito Y (1983) Chinese and African Euphorbiacea plant extracts: Markedly enhancing effect on Epstein-Barr virus-induced transformation. Cancer Lett 19:199–205

Moar MH, Siegert W, Klein G (1978) Autoradiographic detection of Epstein-Barr virus (EBV)-associated early antigen in a variety of EBV DNA containing lymphoblastoid cell lines previously designated as nonproducers. Intervirology 9:333–343

Mondal S, Heidelberger C (1976) Transformation of C3H/10T1/2 CL8 mouse embryo fibroblasts by ultraviolet irradiation and a phorbol ester. Nature 260:710–711

Mondal S, Brankow DW, Heidelberger C (1976) Two-stage chemical oncogenesis in culture of C3H/10T1/2 cells. Cancer Res 36:2254–2260

Morigaki T, Sugawara K, Ito Y (1981) Effect of protease inhibitors on the activation of Epstein-Barr virus repressed in cultured lymphoid cells. Intervirology 16:49–52

Moroney J, Smith A, Tomei D, Wenner CE (1978) Stimulation of $^{86}Rb^{+}$ and ^{32}Pi movements in 3T3 cells by prostaglandins and phorbol esters. J Cell Physiol 95: 287–294

Moss DJ, Pope JH (1972) Assay of the infectivity of Epstein-Barr virus by transformation of human leukocytes in vitro. J Gen Virol 17:233–236

Mueller-Lantzsch N, Yamamoto N, zur Hausen H (1979) Analysis of early and late Epstein-Barr virus associated polypeptides by immunoprecipitation. Virology 97: 378–387

Mueller-Lantzsch N, Georg B, Yamamoto N, zur Hausen H (1980a) Epstein-Barr virus proteins. III. Analysis of polypeptides from P3HR-1-EBV-superinfected NC37 cells by immunoprecipitation. Virology 102:231–233

Mueller-Lantzsch N, Georg B, Yamamoto N, zur Hausen H (1980b) Epstein-Barr virus-induced proteins. II. Analysis of surface polypeptides from EBV-producing and superinfected cells by immunoprecipitation. Virology 102:401–411

Niedel JE, Kuhn LJ, Vandenbark GR (1983) Phorbol diester receptor copurifies with protein kinase C. Proc Natl Acad Sci USA 80:36–40

Nilsson K (1979) The nature of lymphoid cell lines and their relationship to the virus. In: Epstein MA, Achong BG (eds) The Epstein-Barr virus. Springer, Berlin Heidelberg New York, pp 227–281

Niwa O, Sugahara T (1981) 5-azacytidine induction of mouse endogenous type C virus and suppression of DNA methylation. Proc Natl Acad Sci USA 78:6290–6294

Nonoyama M, Pagano JS (1971) Detection of Epstein-Barr viral genome in non-productive cells. Nature 233:103–106

Nonoyama M, Pagano JS (1972a) Replication of viral deoxyribonucleic acid and breakdown of cellular deoxyribonucleic acid in Epstein-Barr virus infection. J Virol 9:714–716

Nonoyama M, Pagano JS (1972b) Separation of Epstein-Barr virus DNA from large chromosomal nonvirus-producing cells. Nature 238:169–171

Nonoyama M, Pagano J (1973) Homology between Epstein-Barr virus DNA and viral DNA from Burkitt's lymphoma and nasopharyngeal carcinoma determined by DNA-DNA reassociation kinetics. Nature 242:44–47
Nonoyama M, Huang CH, Pagano JS, Klein G, Singh S (1973) DNA of Epstein-Barr virus detected in tissue of Burkitt's lymphoma and nasopharyngeal carcinoma. Proc Natl Acad Sci USA 70:3265–3268
O'Brien TG, Lewis MA, Diamond L (1979) Ornithine decarboxylase activity and DNA synthesis after treatment of cells in culture with 12-O-tetradecanoylphorbol-13-acetate. Cancer Res 39:4477–4480
Ohuchi K, Levine L (1978) Stimulation of prostaglandin synthesis by tumor-promoting phorbol-12, 13-diesters in canine kidney (MDCK) cells: cycloheximide inhibits the stimulated prostaglandin synthesis, deacylation of lipids and morphological changes. J Biol Chem 253:4783–4790
Ooka T (1983) DNA polymerase of Epstein-Barr virus (in Japanese). Tanpakushitsu Kakusan Koso 28:271–281
Ooka T, Calender A (1980) Effects of arabinofuranosylthymine on Epstein-Barr virus replication. Virology 104:219–223
Ooka T, Lenoir G, Daillie J (1979) Characterization of an Epstein-Barr virus-induced DNA polymerase. J Virol 29:1–10
Ossowski L, Unkeless JC, Tobia A, Quigley JP, Rifkin DB, Reich E (1973) An enzymatic function associated with transformation of fibroblasts by oncogenic viruses. II. Mammalian fibroblast cultures transformed by DNA and RNA tumor viruses. J Exp Med 137:112–126
Pacifici M, Holtzer H (1977) Effects of tumor-promoting agents on chondrogenesis. Am J Anat 150:207–212
Pearson HE, Lagerborg DL, Winzler RJ (1952) Effects of certain amino acids and related compounds on propagation of mouse encephalomyelitis virus. Proc Soc Exp Biol 79:409–411
Perlmann C, Saemundsen AK, Klein G (1982) A fraction of Epstein-Barr virus virion DNA is methylated in and around the EcoR1-J fragment. Virology 123:217–221
Pietropaolo C, Laskin JD, Weinstein IB (1981) Effect of tumor promoters on sarc gene expression in normal and transformed chick embryo fibroblasts. Cancer Res 41: 1565–1573
Poiesz BJ, Ruscetti FW, Gazdar AF, Bunn PA, Minna JD, Gallo RC (1980) Detection and isolation of type-C retrovirus particles from fresh and cultured lymphocytes of a patient with cutaneous T-cell lymphoma. Proc Natl Acad Sci USA 77:7415–7419
Pope JH, Horne MK, Scott W (1968) Transformation of foetal human leukocytes in vitro by filtrates of a human leukaemic cells line containing herpes-like virus. Int J Cancer 3:857–866
Prtchett R, Pedersen M, Kieff E (1976) Complexity of EBV homologues DNA in continuous lympoblastoid cell lines. Virology 74:227–231
Pulvertaft RTV (1965) A study of malignant tumors in Nigeria by short term tissue culture. J Clin Pathol 18:261–273
Purchio AF, Erikson E, Erikson RL (1977) Translation of 35S and of subgenomic regions of avian sarcoma virus RNA. Proc Natl Acad Sci USA 74:4661–4665
Quigley JP (1979) Phorbol ester-induced morphological changes in transformed chick fibroblasts: evidence for direct catalytic involvement of plasminogen activator. Cell 17:131–141
Razin A, Riggs AD (1980) DNA methylation and gene function. Science 210:604–610
Reedman BM, Klein G (1973) Cellular localization of an Epstein-Barr virus-associated complement-fixing antigen in producer and non-producer lymphoblastoid cell lines. Int J Cancer 11:499–520
Rhim JS, Vass E, Cho HY, Huebner RJ (1971) Malignant transformation induced by 7, 12-dimethyl-benz(a)anthracene in rat embryo cells infected with Rauscher leukemia virus. Int J Cancer 7:65–74

Roberts JW, Roberts CW (1975) Proteolytic cleavage of bacteriolphage lambda repressor in induction. Proc Natl Acad Sci USA 72:147–151

Robinson J, Miller G (1975) Assay for Epstein-Barr virus based on stimulation of DNA synthesis in mixed leukocytes from human umbilical cord blood. J Virol 15:1065–1072

Rose A, Stahn R, Pasovoy DS, Herschman H (1976) Epidermal growth factor enhancement of skin tumor induction in mice. Experientia 32:913

Rovera G, O'Brien TG, Diamond L (1977) Tumor promoters inhibit spontaneous differentiation of Friend erythroleukemia cells in culture. Proc Natl Acad Sci USA 74:2894–2898

Saemundsen AK, Perlmann C, Klein G (1983) Intracellular Epstein-Barr virus DNA is methylated in and around the ecoRl-J fragment in both producer and nonproducer cell lines. Virology 126:701–706

Schneider U, zur Hausen H (1975) Epstein-Barr virus induced transformation of human leukocytes after cell fractionation. Int J Cancer 15:59–66

Seif R (1980) Factors which disorganize microtubules or microfilaments increase the frequency of cell transformation by polyoma virus. J Virol 36:421–428

Shamberger R (1971) Inhibitory effect of vitamin A on carcinogenesis. J Cancer Inst 47:667–673

Shieve W, Skinner CG (1963) Amino acid analogues. In: Hoster RM, Quastel JH (eds) Metabolic inhibitors, vol 1. Academic, New York, pp 2–73

Shin S, Tanaka A, Nonoyama M (1983) Transcription of the Epstein-Barr virus genome in productively infected cells. Virology 124:13–20

Shoyab M, Delarco LE, Todaro GJ (1979) Biologically active phorbol esters specifically alter affinity of epidermal growth factor membrane receptors. Nature 279: 387–391

Shoyab M, Todaro GJ (1980) Specific high affinity cell membrane receptors for biologically active phorbol and ingenol esters. Nature 288:451–455

Sivak A, Ray F, Van Duuren BL (1969) Phorbol ester tumor promoting agents and membrane stability. Cancer Res 29:624–630

Skare J, Farley J (1981) Composition of the transforming EBV recombinant genomes made by cells superinfected with P3HR-1 virus, abstract. Int workshop on herpes viruses, Bologna, pp 242–243

Slaga TJ, Boutwell RK (1978) Mechanisms of tumor promotion and cocarcinogenesis. In: Carcinogenesis, vol 2. Raven, New York

Soprano KJ, Baserga R (1980) Reactivation of ribosomal RNA genes in human-mouse hybrid cells by 12-O-tetradecanoylphorbol 13-acetate. Proc Natl Acad Sci USA 77: 1566–1569

Stastny M, Cohen S (1970) Epidermal growth factor 4. The induction of ornithine decarboxylase. Biochim Biophys Acta 204:578–589

Strauss B, Scudiero D, Henderson E (1974) The nature of the alkylation lesion in mammalian cells. In: Hanawalt PC, Setlow RB (eds) Molecular mechanism for repair of DNA, part A. Plenum, New York, pp 13–24

Strnad BC, Neubauer RH, Robin H, Mazur R (1979) Correlation between Epstein-Barr virus membrane antigen and three large cell surface glycoproteins. J Virol 32:885–894

Sugawara K, Mizuno F, Osato T (1972) Epstein-Barr virus associated antigens in non-producing clones of human lymphoblastoid cell lines. Nature 239:242–243

Sundar SK, Ablashi DV, Armstrong GR, Zipkin M, Faggioni A, Levine P (1981) Steroids inhibit tumor promoting agent induced Epstein-Barr virus early antigens in Raji cells. Int J Cancer 28:503–507

Suss R, Kinzel V, Kreibich G (1971) Cocarcinogenic croton oil factor A1 stimulates lipid synthesis in cell cultures. Experientia 27:46–48

Sutter D, Doerfler W (1980) Methylation of integrated adenovirus type 12 sequences in transformed cells is inversely correlated with viral gene expression. Proc Natl Acad Sci USA 77:253–256

Takai Y (1983) Transmembrane control mechanisms of extracellular signals (in Japanese). Seikagaku 55:226–248
Thorley-Lawson DA, Edson CM (1979) Polypeptides of the Epstein-Barr virus membrane antigen complex. J Virol 32:458–467
Tomkins GM, Levinson BB, Bacter JD, Dethlefsen L (1972) Further evidence for posttranscriptional control of inducible tyrosine aminotransferase synthesis in cultured hepatoma cells. Nature 239:9–14
Tooze J (1981) DNA tumor viruses. In: Molecular biology of tumor viruses, 2nd edn. Cold Spring Harbor Laboratory, New York, pp 1–1073
Touraine JL, Hadden JW, Touraine F, Hadden EM, Estensen R, Good RA (1977) Phorbol myristate acetate: a mitogen selective for a T-lymphocyte population. J Exp Med 145:460–465
Tovey MG, Lenoir G, Begon-Lours J (1978) Activation of latent Epstein-Barr virus by antibody to human IgM. Nature 276:271–273
Toyoshima K, Vogt PK (1969) Temperature sensitive mutants of an avian sarcoma viruses. Virology 39:930–931
Troll W, Klassen A, Janoff A (1970) Tumorigenesis in mouse skin: Inhibition by synthetic inhibitors of proteases. Science 169:1211–1213
Umezawa K, Weinstein IB, Horowitz A, Fujiki H, Matsushima T, Sugimura T (1981) Teleocidin B and the phorbol ester tumor promoters produce similar effects on membranes and membrane receptors. Nature 290:411–413
Unkeless JC, Tobia A, Ossowski L, Quigley JP, Rifkin DB, Reich E (1973) An enzymatic function associated with transformation of fibroblasts by oncogenic viruses. I. Chick embryo fibroblast cultures transformed by avian RNA tumor viruses. J Exp Med 137:85–111
Van Duuren BL (1969) Tumor promoting agents in two-stage carcinogenesis. Prog Exp Tumor Res 11:31–68
Van Duuren BL, Orris L (1965) The tumor enhancing principles of Croton tiglium L. Cancer Res 25:1871–1875
Van Santen V, Cheung A, Kieff E (1981) Epstein-Barr virus RNA 7. Viral cytoplasmic RNA in a restringently infected cell line transformed in vitro by EBV. Proc Natl Acad Sci USA 78:1930–1934
Verma AK, Boutwell RK (1977) Vitamin A acid (retinoic acid), a potent inhibitor of 12-O-tetradecanoylphorbol-13-acetate-induced ornithine decarboxylase activity in mouse epidermis. Cancer Res 37:2196–2201
Vogt PK (1977) Genetics of RNA tumor viruses. In: Fraenkel-Conrat H, Wagner RR (eds) Comprehensive virology, vol 9. Plenum, New York, pp 401–483
Weigel R, Miller G (1983) Major EB virus-specific cytoplasmic transcripts in a cellular clone of the HR-1 Burkitt lymphoma line during latency and after induction of viral replicative cycle by phorbol esters. Virology 125:287–298
Weinstein IB (1983) Molecular mechanisms of multistage carcinogenesis, abstract. 14th Int symp of The Princess Takamatsu Cancer Res Fund on cellular interaction by environmental tumor promoters, pp 14–15
Weinstein IB, Wigler M, Fisher PB, Sisskin E, Pietropaolo C (1978) Cell culture studies on the biologic effects of tumor promoters. In: Slaga TJ, Sivak A, Boutwell RK (eds) Mechanisms of tumor promotion and cocarcinogenesis, vol 2. Raven, New York, pp 313–333
Weinstein IB, Lee LS, Fisher PB, Mufson RA, Yamasaki H (1979) The mechanism of action of tumor promoters and a model of two stage carcinogenesis. In: Emmelot P, Kriek E (eds) Environmental carcinogenesis. Elsevier, Amsterdam, pp 265–285
Weissbach A, Hong SCL, Auker J, Muller R (1973) Characterization of herpes simplex virus-induced deoxyribonucleic acid polymerase. J Biol Chem 248:6270–6277
Wigler M, Weinstein IB (1976) Tumor promoter induces plasminogen activator. Nature 259:232–233
Wolf H, zur Hausen H, Becker V (1973) EB viral genomes in epithelial nasopharyngeal carcinoma cells. Nature 244:245–247

Wunderlich V, Baumbach L, Sydow G (1982) Tumor promoter-stimulated synthesis of Mason-Pfizer monkey virus. Cancer Res Clin Oncol 102:271–275

Yamamoto H, Katsuki T, Harada S, Hinuma Y (1981a) Enhancement of outgrowth of EB virus-transformed cells from normal human peripheral blood by a tumor promoter, TPA. Int J Cancer 27:161–166

Yamamoto H, Katsuki T, Hinuma Y, Hoshino H, Miwa M, Fujiki H, Sugimura T (1981b) Induction of Epstein-Barr virus by a new tumor promoter, teleocidin, compared to induction by TPA. Int J Cancer 28:125–129

Yamamoto N, Bauer G (1981) The relationship between tumor promoter binding and Epstein-Barr virus induction in human lymphoblastoid cell lines. J Cell Physiol 109: 397–402

Yamamoto N, Hinuma Y (1976) Clonal transformation of human leukocytes by Epstein-Barr virus in soft agar. Int J Cancer 17:191–196

Yamamoto N, Hinuma Y (1978) Induction of nuclear antigen (EBNA) synthesis without transformation in human cord lymphocytes by P3HR-1 strain of Epstein-Barr virus. Virology 89:623–626

Yamamoto N, zur Hausen H (1979) Tumor promoter TPA enhances transformation of human leukocytes by Epstein-Barr virus. Nature 280:244–245

Yamamoto N, zur Hausen H (1980) Effect of inhibition of DNA synthesis on Epstein-Barr virus induction by tumor promoters. Virology 101:104–110

Yamamoto N, zur Hausen H (1981) Induction of Epstein-Barr virus early antigens by intercalating chemicals in B95-8 cells. Virology 115:390–394

Yamamoto N, Katsuki T, Hinuma Y (1976) Transformation of tonsil lymphocytes by Epstein-Barr virus. J Natl Cancer Inst 56:1105–1107

Yamamoto N, Bister K, zur Hausen H (1979) Inhibition of Epstein-Barr virus induction by retinoic acid. Nature 278:553–554

Yamamoto N, Mueller-Lantzsch N, zur Hausen H (1980a) Differential inhibition of Epstein-Barr virus induction by the amino acid analogue L-canavanine. Int J Cancer 25:439–443

Yamamoto N, Mueller-Lantzsch N, zur Hausen H (1980b) Effect of actinomycin D and cycloheximide on Epstein-Barr virus early antigen induction in lymphoblastoid cells. J Gen Virol 51:255–261

Yamamoto N, Okada M, Koyanagi Y, Kannagi M, Hinuma Y (1982) Transformation of human leukocytes by cocultivation with an adult T-cell leukemia virus (ATLV) producer cell line. Science 217:737–739

Yamasaki H, Fibach E, Nudel U, Weinstein IB, Rifkind RA, Marks PA (1977) Tumor promoters inhibit spontaneous and induced differentiation of murine erythroleukemia cells in culture. Proc Natl Acad Sci USA 74:3451–3455

Yamasaki H, Weinstein IB, Fibach E, Rifkind RA, Marks PA (1979) Tumor promoter induced adhesion of the DS 19 clone of murine erythroleukemia cells. Cancer Res 39:1989–1994

Yoshida M, Miyoshi I, Hinuma Y (1982) Isolation and characterization of human adult T-cell leukemia and its implication in the disease. Proc Natl Acad Sci USA 79: 2031–2035

Yotti LP, Chang CC, Trosko JE (1979) Elimination of metabolic cooperation in Chinese hamster cells by a tumor promoter. Science 206:1089–1091

Youssoufian H, Hammer SM, Hirsch MS, Mulder C (1982) Methylation of the viral genome in an in vitro model of herpes simplex virus latency. Proc Natl Acad Sci USA 79:2207–2210

Yuspa SH, Lichti U, Ben T, Patterson E, Hennings H, Slaga TJ, Colburn N, Kelsey W (1976) Phorbol-ester tumor promoters stimulate DNA synthesis and ornithine decarboxylase activity in mouse epidermal cell cultures. Nature 262:402–404

Zucker MB, Troll W, Belman S (1974) The tumor promoter phorbol ester (12-O-tetradecanoylphorbol-13-acetate), a potent aggregating agent for blood platelets. J Cell Biol 60:325

Zeng Y, Xhou HM, Xu SP (1981) Inhibitory effect of retinoids on Epstein-Barr virus induction in Raji cells. Intervirology 16:29–32
zur Hausen H (1975) Oncogenic herpes viruses. Biochim Biophys Acta 417:25–53
zur Hausen H (1980) The role of viruses in human tumors. Adv Cancer Res 32:77–107
zur Hausen H, Fresen KO (1977) Heterogeneity of Epstein-Barr virus. II. Induction of early antigen (EA) by complementation. Virology 81:138–143
zur Hausen H, Schulte-Holthausen H (1970) Presence of EB virus nucleic acid homology in a "virus-free" line of Burkitt tumor cells. Nature 227:245–248
zur Hausen H, Henle W, Hummeler K, Diehl V, Henle G (1967) Comparative study of cultured Burkitt tumor cells by immunofluorescence, autoradiography, and electron microscopy. J Virol 1:830–837
zur Hausen H, Schulte-Holthausen H, Klein G, Henle W, Henle G, Clifford P, Santesson L (1970) EBV DNA in biopsies of Burkitt tumors and anaplastic carcinomas of the nasopharynx. Nature 228:1056–1058
zur Hausen H, Diehl V, Wolf H, Schulte-Holthausen H, Schneider U (1972) Occurrence of EB virus genomes in human lymphoblastoid cell lines. Nature 237:189–190
zur Hausen H, O'Neill FJ, Freeze JK, Hecker E (1978) Persisting oncogenic herpes virus induced by the tumor promoter TPA. Nature 272:373–375
zur Hausen H, Bornkamm GW, Schmidt R, Hecker E (1979) Tumor initiators and promoters in the induction of Epstein-Barr virus. Proc Natl Acad Sci USA 76:782–785
zur Hausen H, Yamamoto N, Bauer G (1982) Virus induction by tumor promoters. In: Hecker E, Fusenig NE, Kunz W, Marks F, Thielmann HW (eds) Cocarcinogenesis and biological effects of tumor promoters. Carcinogenesis, vol 7, pp 617–623

Rev. Physiol. Biochem. Pharmacol., Vol. 101

Factors Involved in the Physiological Regulation of the Cerebral Circulation

DAVID W. BUSIJA[1] and DONALD D. HEISTAD[2]

Contents

---

1 Department of Physiology and Biophysics, University of Tennessee Center for the Health Sciences, Memphis, Tennessee 38163, USA
2 VA Medical Center and Department of Internal Medicine, University of Iowa College of Medicine and Cardiovascular Center, Iowa City, Iowa 52242, USA

1 Introduction

Blood flow in the cerebral circulation, as in other regional circulations, is regulated by four primary factors; metabolic stimuli, perfusion pressure, chemical stimuli, and autonomic nerves. However, the relative importance of each of these factors in the cerebral circulation differs from that in other circulations. We will review recent studies concerning the role played by each of these factors in the regulation of cerebral blood flow (CBF) and cerebrovascular resistance (CVR).

2 Vascular Anatomy of the Brain

The brain is supplied primarily by two sets of paired arteries, the carotid and vertebral arterial systems (*Carpenter* 1978). The vertebral arteries fuse at the junction of the pons and medulla oblongata and form the basilar artery. The internal carotid arteries and the basilar artery join to form the circle of Willis at the base of the cerebrum. From this vascular backbone, projecting tributaries of these arteries supply the brain parenchyma. In general, the carotid arterial system supplies rostral areas of the brain (cerebrum and diencephalon) while the vertebral system supplies caudal areas of the brain (midbrain, pons, medulla, and cerebellum) (*Wellens* et al. 1977; *Heistad* et al. 1980a; *Orr* et al. 1983). The venous drainage of the brain is more variable and differs among species (*Hegedus* and *Shackelford* 1965). Although it was previously believed that the brain does not have a well-defined lymphatic system compared with other organs, approximately 50% of albumin injected into cerebral tissues passes into deep cervical lymphatics (*Bradbury* et al. 1981). In addition, the cerebrospinal fluid also is involved in removal of proteins from the brain interstitial space (*Davson* 1967; *Bradbury* et al. 1981).

Two unique features of the cerebral circulation deserve comment. First, in apparent contrast to other regional circulations, large arteries, including internal carotid and basilar arteries, circle of Willis, and direct branches of these arteries, are important resistance vessels (*Heistad* et al. 1978a; *Kontos* et al. 1978a; *Baumbach* and *Heistad* 1983). It has been suggested that this arrangement protects cerebral microvessels and safeguards the cerebral circulation from vascular "steal" during focal increases in blood flow in the brain (*Abboud* 1981). Large cerebral arteries and arterioles may form a pressure equalization reservoir, which maintains uniform pressure over the brain surface, while smaller arterioles serve as distributors of blood flow to meet local conditions. Second, brain blood vessels, especially the small arterioles and capillaries, possess a morphological and biochemical barrier between blood and brain interstitial space. The morphological barrier is composed of continuous endothelium, with tight junctions between adjacent cells (*Rapoport* 1976). In addition, there are few canaliculi through the endothelium and little evidence of important vesicular transport across the cells (*Rapoport* 1976; *Edvinsson* and *MacKenzie* 1977). The biochemical barrier consists of high levels of degradative enzymes that serve to prevent circulating neurotransmitters from passing from blood to brain (*Rapoport* 1976; *Edvinsson* and *MacKenzie* 1977; *Hardebo* and *Owman* 1980).

Oldendorf (1971) has compared the penetration of ^{14}C-amines with a "freely diffusible" test substance (^{3}HOH) and found that the "brain-uptake index" for neurotransmitters, such as norepinephrine, was only approximately 4%. The blood-brain barrier is not absolute: In some areas of the brain, such as the posterior pituitary, median eminence, and area postrema, the barrier is less restrictive (*Rapoport* 1976). The barrier may be disrupted transiently by hyperosmolar substances (*Rapoport* 1976) or acute hypertension (*Johansson* and *Norborg* 1978). Under the latter conditions, circulating neurotransmitters that are normally excluded from cerebral vascular smooth muscle and the central nervous system may affect vascular tone directly or indirectly via changes in cerebral metabolism. For example, norepinephrine, which normally has minimal effects on CBF when infused intra-arterially, increases CBF following osmotic opening of the blood-brain barrier, largely by increasing cerebral metabolic rate (*MacKenzie* et al. 1976a). In contrast, intracarotid infusion of histamine under similar conditions increases CBF exclusively via vascular H_1 and H_2 receptors (*Gross* et al. 1981a). Although histamine does not increase CBF when the blood-brain barrier is intact, it does increase cerebral vascular permeability modestly (*Gross* et al. 1982).

Neurotransmitters normally do not transverse the blood-brain barrier, but antagonists, precursors, and related agonists often do cross the barrier. For example, dopamine crosses the blood-brain barrier poorly (*Oldendorf* 1971) but agonists (e.g., apomorphine) (*McCulloch* et al. 1982a) and antagonists (e.g., haloperidol) (*McCulloch* et al. 1982b) transverse the blood-brain barrier and have neurological effects. In addition, L-dopa readily crosses the blood-brain barrier and is rapidly converted into dopamine (*Oldendorf* 1971; *De la Torre* and *Mullan* 1971).

3 Methods to Characterize Cerebrovascular Responses

Understanding of the regulation of the cerebral circulation has been hindered seriously in comparison with other regional circulations because of anatomic complexities of the vascular system. Classic methods, such as measurement of arterial inflow and venous outflow, are not optimal for studies of brain circulation. For example, the internal carotid artery, even in primates, supplies extracerebral tissues in addition to the brain, and supplies heterogeneous and variable areas within the brain (*Edvinsson* and *MacKenzie* 1977; *Marcus* et al. 1981a). In addition, in some species, such as dogs, there are such extensive anastomoses between branches of the internal and external carotid arteries that infusion of vasodilator drugs directly into the internal carotid increases blood flow to extracerebral structures, but not to the brain (*Heistad* et al. 1980a). Further, except for the venous outflow method of *Rapela* and *Green* (1964; *Traystman* and *Rapela* 1975), measurement of outflow through veins draining the head suffers from similar liabilities (*Edvinsson* and *MacKenzie* 1977).

Recently, we and others have reviewed methods to measure CBF (*Edvinsson* and *MacKenzie* 1977; *Heistad* et al. 1978; *Marcus* et al. 1981b). We suggest that methods using clearance of inert gases such as ^{133}Xe and hydrogen, and heat clearance, suffer from serious drawbacks which restrict their usefulness in the measurement of CBF (*Edvinsson* and *MacKenzie* 1977; *Marcus* et al. 1981a). However, we have suggested that the ^{133}Xe clearance method is valid when CBF is less than 120 ml/min/100 g, when ^{133}Xe is given by intracarotid injection, and when stochastic analysis is used (*Marcus* et al. 1981b).

There are several methods that allow accurate characterization of cerebral vascular responses. First, visual observation and measurement of pial vessels with either the closed (*Levasseur* et al. 1975) or open window (*Kuschinsky* et al. 1974) preparations allow virtually continuous examination of the effects of physiological and pharmacological stimuli in anesthetized animals. Topical application of drugs circumvents the blood-brain barrier. However, limitations are that only one segment of the total circulation is exposed and vascular diameter, not blood flow, is measured. The latter limitation is overcome by concurrent measurement of blood flow velocity. Blood flow velocity has been measured simultaneously with diameter using a miniaturized pulsed-doppler probe (*Busija* et al. 1981) or timed passage of a saline pulse (*Gotoh* et al. 1982). Calculated blood flow using the former method has been compared with measurements with microspheres and the agreement has been excellent (*Busija* et al. 1981). Measurement of intravascular pressure also has been combined with diameter measurement. Second, the use of diffusible indicator substances, such as ^{14}C-iodoantipyrine, allows accurate measurement of CBF with excellent spatial resolution in anesthetized and awake animals (*Sakurada* et al. 1978). A drawback of this method is that only one determination per animal is possible. Third, the cerebral venous outflow method of *Rapela* and *Green* (1964) allows continuous monitoring of blood flow draining the dorsal cerebrum. Although there appears to be no significant extracerebral contamination, it is unclear precisely which areas of the brain are drained, and the method cannot be used in awake animals. A venous outflow method also has been developed for use in rats (*Nilsson* et al. 1979). Fourth, the radioactive microsphere method allows multiple (up to six) determinations of regional and total CBF in anesthetized and awake animals, as well as simultaneous determination of other regional blood flows. We have examined this method and have shown that it can measure CBF accurately (*Marcus* et al. 1976, 1981a; *Busija* et al. 1981, 1982b). In addition, the agreement of regional blood flows determined simultaneously with microspheres and ^{14}C-iodoantipyrine is excellent (*Schuier* et al. 1981). A drawback of the microsphere method is that it does not provide continuous measurement of CBF. However, simultaneous measurement of vascular diameters and/or blood flow velocity, or cerebral venous outflow, largely circumvents this limitation.

4 Chemical Stimuli

4.1 Arterial PCO_2 ($PaCO_2$)

Alterations in $PaCO_2$ profoundly affect cerebral vascular tone. Hypocapnia results in arterial constriction and reduced CBF (*Raper* et al. 1971; *Busija* et al. 1981), whereas hypercapnia causes arterial dilatation and increased CBF (*Raper* et al. 1971; *Busija* and *Heistad* 1981). Cerebrovascular changes are due solely to changes in $PaCO_2$, as alterations in arterial pH independent of $PaCO_2$ have minimal effects on CBF (*Harper* and *Bell* 1963). During hypercapnia, CBF increases in all major regions of the brain, although the magnitude of this change is greater in cortical gray matter than in white matter (*Busija* and *Heistad* 1981). Hypercapnia reduces cerebral glucose utilization (*Des Rosiers* et al. 1978; *Miller* et al. 1975). A recent preliminary report suggests that, in contrast to other regions of the brain, blood flow to the choroid plexus and posterior pituitary increases minimally during hypercapnia (*Hanley* et al. 1983).

The temporal response of blood flow to abrupt increases in inspired CO_2 is rapid, such that detectable changes in pial arterial diameter or CBF are apparent within 1–2 min and steady-state responses are reached within 8–12 min (*Raper* et al. 1971; *Busija* et al. 1981). Resistance vessels of all sizes contribute to the increase in CBF during hypercapnia, including circle of Willis and basilar artery (*Heistad* et al. 1978a), pial arteries (*Raper* et al. 1971), and parenchymal arteries (*Sadoshima* et al. 1980). Atherosclerosis impairs maximal cerebral vasodilator responses to hypercapnia, possibly by affecting primarily large arteries (*Heistad* et al. 1980b). Cerebral vasodilatation during hypercapnia occurs in all age-groups, although there are modest developmental differences. When normalized by cerebral metabolic rate for oxygen, the increase in CBF in unanesthetized sheep is greatest in newborns, smaller in adults, and even smaller in fetuses (*Rosenberg* et al. 1982).

Several mechanisms have been proposed to account for the relationship between $PaCO_2$ and cerebrovascular tone. These mechanisms include neural pathways, prostaglandins, and extracellular fluid (ECF) $[H^+]$ in the brain.

Several investigators have suggested that hypercapnia activates cholinergic reflex pathways to cerebral vessels which increase CBF (*James* et al. 1969; *Rovere* et al. 1973; *Shalit* et al. 1967). However, recent findings demonstrate that denervation of peripheral chemoreceptors (*Heistad* et al. 1976), section of the greater superficial petrosal nerve (*Busija* and *Heistad* 1981), or administration of atropine (*Busija* and *Heistad* 1982) do not alter the response to hypercapnia. In addition, *Kontos* et al. (1977b) found that flushing the cortical surface with CSF with reduced PCO_2 prevented pial arterial dilatation during systemic hypercapnia. Thus, recent evidence suggests that reflex pathways do not contribute substantially to cerebral vasodilatation during hypercapnia. In contrast, activation of sympathetic constrictor pathways during hypercapnia appears to restrict vasodilatation (*Wei* et al. 1980c; *Busija* and *Heistad* 1984).

Prostaglandins synthesized by cerebral vessels or tissues have also been suggested as mediators linking $PaCO_2$ and CBF. Numerous investigators have presented evidence which indicates that inhibition of cyclo-oxygenase activity by indomethacin, in particular, reduces CBF by 30%–50% during normocapnia and virtually abolishes the increase in CBF during hypercapnia (*Pickard* and *MacKenzie* 1973; *Bill* 1979; *Sakabe* and *Siesjö* 1979; *Dahlgren* et al. 1981a,c; *Crockard* et al. 1982; *McCulloch* et al. 1982d) without affecting cerebral metabolism (*McCulloch* et al. 1982d). None of these studies, however, has presented direct evidence indicating that prostaglandin synthesis in cerebral tissues or vessels was inhibited, or that inhibition of prostaglandin synthesis correlates with reduction of response to CO_2, and only two studies (*Pickard* and *MacKenzie* 1973; *Sakabe* and *Siesjö* 1979) have determined whether indomethacin depresses responses of cerebral vessels in a nonspecific fashion. Indomethacin may be inactivated by improper preparation and may inhibit other enzyme systems in addition to cyclo-oxygenases (*Flower* 1974). In addition, the vehicle used for indomethacin may affect cerebral vascular responses independent of actions of the drug. For example, *Wei* et al. (1980b) found that prolonged infusion into cerebral arteries of a solution similar to the vehicle used by *Pickard* and *MacKenzie* (1973) attenuated cerebrovascular dilatation during hypercapnia. In addition, *Crockard* et al. (1982) found that imidazole, as well as indomethacin, prepared and infused in a similar way, blocked the increase in CBF during hypercapnia, even though imidazole typically does not block prostaglandin synthesis.

In contrast, we (*Busija* and *Heistad* 1983a; *Busija* 1983a) and others (*Wei* et al. 1980b; *Jackson* et al. 1983) have reported that indomethacin, at doses that effectively attenuate prostaglandin synthesis of cerebral vessels and/or tissues without having nonspecific depressant effects on cerebral vessels, does not affect CBF or pial arterial diameter in anesthetized cats or CBF in awake rabbits or anesthetized dogs. The cerebrovascular response to indomethacin may not necessarily be related to inhibition of prostaglandin synthesis, but could rather be due to a direct, nonspecific effect of indomethacin on cerebral vessels. For example, *Eriksson* et al. (1983) found that while acute administration of indomethacin affected CBF in awake man, acute administration of aspirin and chronic administration of aspirin and indomethacin did not decrease resting CBF or attenuate the increase in CBF during hypercapnia. Cyclo-oxygenase inhibitors other than indomethacin do not appear to inhibit responses to hypercapnia. *Wei* et al. (1980b) found that AHR-5850 (Sodium Anfenac) did not affect pial arterial diameter during normocapnia or hypercapnia. In addition, *Pickard* et al. (1977) and *Amano* and *Meyer* (1981) reported that aspirin did not affect CBF during hypercapnia in awake man.

It is possible that species differences may account for divergent findings. For example, indomethacin may affect CBF in rats (*Sakbe* and *Siesjö* 1979; *Dahlgren* et al. 1981a,c; *McCulloch* et al. 1982d), nonhuman primates (*Pickard* and *MacKenzie* 1973), and man (*Eriksson* et al. 1983) by a mechanism unrelated to inhibition of prostaglandin synthesis.

Most of the evidence concerning the role of prostaglandins in the regulation of CBF has been pharmacological. Recently, several laboratories have reported on biochemical determinations of prostaglandins in the brain or cerebral venous blood during changes in $PaCO_2$. *Ellis* et al. (1982) found that prostaglandin E_2 and $F_{2\alpha}$ levels in brain tissue actually decreased from normocapnia levels when cats were exposed to hypercapnia, while 6-keto-$F_{1\alpha}$ did not change, and pial arterial diameter increased 32%. In addition, *McCalden* et al. (1982), *Eriksson* et al. (1983), and *Jackson* et al. (1983) found that prostacyclin levels in cerebral venous blood in baboons, man, and dogs did not increase from control levels during hypercapnia, while CBF increased. Thus, biochemical determinations of prostaglandins during hypercapnia suggest that prostaglandins do not play an important role in cerebral vasodilator responses to hypercapnia.

Current evidence supports the concept that $PaCO_2$ affects cerebrovascular tone via changes in brain ECF $[H^+]$. Carbon dioxide readily diffuses from blood to brain and causes alterations in ECF $[H^+]$. This mechanism was originally proposed by *Gotoh* et al. (1961), *Skinhøj* (1966), and *Lassen* (1968). Varying CSF pH in the vicinity of pial arteries directly alters pial arterial diameter, so that acidosis dilates and alkalosis constricts vessels (*Kuschinsky* et al. 1972; *Schneider* et al. 1977). *Kontos* et al.

(1977a,b) found that changes in CSF [HCO_3^-] and PCO_2, independent of changes in pH, had no effect on pial arterial tone; in contrast, pial arterial diameter was directly related to CSF pH. They also found that during systemic hypercapnia, flushing the surface of the brain with CSF containing a lower PCO_2 reversed the pial arterial dilatation. Ventriculocisternal perfusion with CSF containing high [HCO_3^-] greatly attenuates the increase in CBF during hypercapnia, but not during hypoxia, which suggests that the mechanism influencing CBF during hypercapnia is acidosis but the mechanism during hypoxia is not predominantly acidosis (*Koehler* and *Traystman* 1982).

During hypocapnia, CBF decreases, due to an increase in CSF pH. However, when $PaCO_2$ is reduced below 20 mmHg, CBF falls no further (*Busija* and *Heistad* 1981), perhaps due to tissue hypoxia and a resultant increase in brain adenosine concentration (*Rubio* et al. 1975; *Winn* et al. 1981), which may have opposing effects on the tone of vascular smooth muscle.

4.2 Arterial PO_2 (PaO_2)

Alterations in PaO_2 have major effects on cerebrovascular tone. Reduction of PaO_2 below 40–50 mmHg results in arterial dilatation and increased CBF (*Borgström* et al. 1975; *Kogure* et al. 1970a; *Kontos* et al. 1978a; *Busija* and *Heistad* 1981). There is an almost linear relationship between arterial blood O_2 content (CaO_2) and cerebrovascular tone (*Borgström* et al. 1975; *Jones* et al. 1981). Because of the shape of the hemoglobin dissociation curve, CaO_2 does not fall appreciably until PaO_2 decreases below 40–50 mmHg. The increase in CBF during hypoxia is sufficient to maintain O_2 delivery constant until CaO_2 falls to very low levels (<4 vol.%), as cerebral O_2 extraction remains unchanged up to this level of hypoxia (*Jones* et al. 1981). Cerebral vasodilatation during hypoxia is present in fetuses, newborns, and adults (*Jones* et al. 1978, 1981; *Busija* and *Heistad* 1981). During hypoxia, blood flow increases to all major areas of the brain, although the increase is greater to cortical gray matter than to white matter (*Busija* and *Heistad* 1981). The temporal course of cerebrovascular dilatation to hypoxia is rapid, such that detectable changes in pial arterial diameter and CBF occur within 30–60 s (*Nilsson* et al. 1975; *Craigen* and *Jennett* 1981; *Busija* and *Heistad* 1981), and steady-state responses are achieved within 5–10 min (*Borgström* et al. 1975).

Increases in PaO_2 at normal atmospheric pressure have little or no effect on pial arterial tone or CBF when $PaCO_2$ is maintained constant (*Kontos* et al. 1978b; *Busija* et al. 1980b). During inhalation of 100% O_2 at 3.5 atm, however, CBF falls substantially, probably due to accompanying hypocapnia (*Lambertsen* et al. 1953).

Several mechanisms have been proposed to account for the relationship between PaO_2 and cerebrovascular tone during hypoxia. Although the average level of brain metabolism does not change during hypoxia, it does increase in several areas, including cerebellum, cerebral white matter, subthalamic, interpendicular, and caudate nuclei (*Traystman* and *Fitzgerald* 1981; *Weiss* et al. 1983; *Miyaoka* et al. 1980; *Duffy* et al. 1982). Thus, part of the cerebral vasodilatation in several areas during hypoxia may be due to an increase in metabolism, but other mechanisms are undoubtedly more important in the overall response. Proposed mechanisms include activation of neural pathways and H^+, K^+, and adenosine concentrations.

Several investigators have suggested that hypoxia activates reflex pathways to cerebral vessels, thereby increasing CBF independent of local mechanisms (*Ponte* and *Purves* 1974; *James* and *MacDonnell* 1975). However, recent findings demonstrate that denervation of peripheral chemoreceptors (*Heistad* et al. 1976; *Traystman* and *Fitzgerald* 1981) and section of the greater superficial petrosal nerve (*Busija* and *Heistad* 1981) do not attenuate the vasodilator response. In addition, *Kontos* et al. (1978b) have shown that superfusion of the cortical surface with CSF or fluorocarbons containing O_2 prevents pial arterial dilatation due to systemic hypoxia. Thus, local mechanisms are sufficient to account for cerebral vasodilator responses to hypoxia, and reflex pathways apparently do not contribute substantially to cerebral vasodilatation during hypoxia. In addition, hypoxia activates sympathetic constrictor pathways which limit cerebral vasodilatation (*Vatner* et al. 1980; *Busija* 1984a).

Systemic hypoxia can lead to production of lactic acid by the brain and it was thought that the resultant acidosis leads to increased CBF (*Betz* 1972). However, recent studies indicate that cerebral acidosis is not an important factor in cerebral vascular responses during the initial period of hypoxia, and may not be sufficient to maintain increased CBF even during prolonged hypoxia. During the first 60 s of hypoxia, as CBF begins to rise, brain lactic acid concentration increases only modestly (*Nordberg* and *Siesjö* 1975), and ECF pH does not change (*Astrup* et al. 1976). Although lactic acid concentration in the brain increases during prolonged hypoxia, this increase appears insufficient to account for the level of CBF (*Nordberg* and *Siesjö* 1975; *Rubio* et al. 1975). Experiments using ventriculocisternal perfusion with CSF containing high [HCO_3^-] also provide evidence that acidosis is only a minor factor during hypoxia. Although perfusion with CSF containing high [HCO_3^-] substantially attenuates the increase in caudate nucleus blood flow during hypercapnia, similar experiments during hypoxia show that blood flow to the caudate nucleus is not reduced appreciably (*Koehler* and *Traystman* 1982). However, *Astrup* et al. (1979) report that ECF pH falls by 0.03 after 20 min of hypoxia, which suggests that acidosis may be involved in the vasodilatation during prolonged hypoxia. In general, current evidence suggests that the contribution of cerebral acidosis to the increase in CBF during hypoxia is relatively minor.

ECF [K^+] also has been implicated in the cerebral hyperemia during hypoxia. ECF [K^+] increases during the 1st min of hypoxia, while ECF pH does not change in the appropriate direction (*Silver* 1973; *Astrup* et al. 1976; *Morris* 1974; *Krishner* et al. 1975, 1976). Topical application of CSF that contains increasing [K^+] dilates pial arteries progressively in the range 0–10 mmol (*Kuschinsky* et al. 1972), presumably via a mechanism involving the electrogenic pump (*Harder* 1980). However, the increase in ECF [K^+] does not appear to be sufficient to account for the entire increase in CBF during the initial period of hypoxia, especially because glial cells would be expected to limit increases in ECF [K^+] (*Orkland* et al. 1966).

Adenosine concentration in the ECF also has been suggested as a factor leading to increased CBF during hypoxia. Adenosine is produced by the dephosphorylation of AMP by 5′-nucleotidase, an enzyme apparently confined to the wall of the glial foot plate and not associated with neurons (*Kreutzberg* et al. 1978). Thus, it has been suggested that the majority of brain adenosine is located extracellularly. However, recent evidence indicates that in the heart a large proportion of adenosine is synthesized and stored within cells (*Olsson* et al. 1982; *Schrader* 1983). At the onset of hypoxia, at a time when CBF increases substantially, brain adenosine levels increase sixfold (*Winn* et al. 1981). In addition, within 5 s of the onset of brain ischemia due to reduction of arterial pressure to zero, brain adenosine concentration increases 2.5-fold (*Winn* et al. 1979). Adenosine is a potent dilator of pial arteries (*Wahl* and *Kuschinsky* 1976) and increases CBF when infused intra-arterially (*Heistad* et al. 1981). Adenosine causes relaxation of vascular smooth muscle by blocking the entry of calcium into the cells (*Berne* 1980). Pharmacological studies support the concept that increased brain adenosine levels lead to cerebral hyperemia during hypoxia. A number of investigators have found that intravenous theophylline, a competitive inhibitor of adenosine, markedly attenuates the increase in CBF during hypoxia (*Emerson* and *Raymond* 1981; *Morii* et al. 1983). In addition, *Wei* and *Kontos* (1983) found that topical application of adenosine deaminase in CSF greatly attenuated cerebral vasodilatation during hypoxia. Thus, recent evidence suggests that adenosine is an important mediator of cerebral vasodilatation during hypoxia.

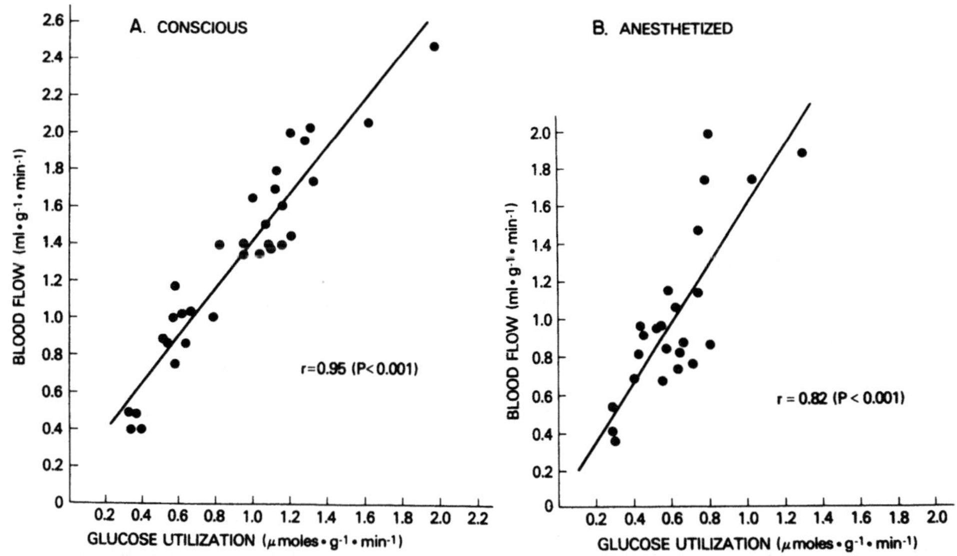

Fig. 1. Relationship between local cerebral blood flow (^{14}C-iodoantipyrine technique) and local cerebral utilization of glucose (^{14}C-deoxyglucose technique) in conscious and anesthetized (Thiopental) rats. Each *point* represents blood flow and glucose utilization in different structures of the brain. (After *Sokoloff* 1981)

5 Metabolism

5.1 Relationship Between Cerebral Metabolism and Blood Flow

Numerous studies have demonstrated a close correlation between neuronal activity and blood flow in the central nervous system. Functional activity of specific brain areas by sensory input or exercise is accompanied by increases in local or regional blood flow (*Kato* et al. 1974; *Leninger-Follert* and *Hossmann* 1979; *Olesen* 1971; *Lassen* et al. 1978; *Foreman* et al. 1976; *Gross* et al. 1980). In addition, electrical stimulation of the cortex and seizures induced with pharmacological agents also increase CBF (*Heuser* 1979; *Mueller* et al. 1979). Quantitation of this relationship in small areas of the brain was made possible by the development of the ^{14}C-iodoantipyrine method for determining CBF (*Sakurada* et al. 1978) and the ^{14}C-deoxyglucose method for determining brain glucose utilization (*Sokoloff* et al. 1977). Using these methods, it was found that the correlation between regional cerebral blood flow and metabolism is excellent in awake and anesthetized animals during normal conditions (*Sokoloff* 1981, Fig. 1), during other conditions, such as metabolic acidosis (*Kuschinsky* et al 1981), and following administration of apomorphine (*McCulloch* et al. 1982c). In addition, in man, there is a good correlation between local O_2 consumption by the $^{15}O_2$ method and blood flow in the same region (*Raichle* et al. 1976). It appears that resting energy metabolism is primarily devoted to such cellular processes as cell maintenance and repair, axoplasmic transport, and protein synthesis, while the increase in energy metabolism with stimulation is a function of the sodium pump (*Mata* et al. 1980).

5.2 Coupling Mechanisms

Cerebral metabolism and blood flow appear to be linked by the release of vasodilator substances from active neurons. These substances reach cerebral resistance vessels by diffusion, dilate the arteries, and increase CBF. The question of which factors couple metabolism and blood flow in the brain has been controversial. Suggested mechanisms include ECF $[H^+]$, $[K^+]$, and adenosine. Usually, correlations between these substances and CBF have been investigated during neuronal activation with seizures or by electrical stimulation. Consequently, coupling between metabolism and CBF is investigated during the period when energy metabolism is increased because of augmented sodium pump activity.

During activation of neurons and increases in metabolism, both increased O_2 consumption and CO_2 production could lead to local acidosis. For example, if O_2 is consumed faster than it is delivered, lactic acid could accumulate. In addition, production of CO_2 by oxidative phosphorylation could lead to an increase in local $[H^+]$. The hydrogen ion is a potent dilator of cerebral arteries and, as mentioned previously, it is largely responsible for the cerebrovascular dilatation during hypercapnia. During electrical stimulation of the cortex, cortical pH has been found to decrease at a time when CBF increases (*Heuser* 1979), or initially to increase and then decrease during prolonged stimulation (*Urbanics* et al. 1978). During seizures, lactic acid accumulates in the extracellular space (*Chapman* et al. 1977). However, measurements of pH reveal that ECF is slightly alkalotic or does not change (*Astrup* et al. 1979; *Urbanics* et al. 1978; *Meldrum* and *Nilsson* 1976), or is acidotic (*Heuser* 1979) during the initial phase of increased CBF. Part of the increase in CBF during seizures is due to severe arterial hypertension which may mask the role of hydrogen or potassium ions. In experiments in which arterial blood pressure is controlled during seizures, perivascular pH begins to fall immediately at a time when pial arteries are dilating (*Kuschinsky* and *Wahl* 1979). Thus, it is likely that ECF $[H^+]$ influences CBF during seizures and perhaps during the later phases of electrical stimulation of the cortex.

It is unclear whether tissue PO_2 actually decreases during activation of neurons. For example, it has been reported that PO_2 decreases (*Ingvar* and *Lassen* 1962; *Metzger* 1979; *Caspers* and *Speckmann* 1972), remains constant (*Silver* 1978), and increases (*Leninger-Follert* and *Lübbers* 1976) during electrical and chemical stimulation of the brain. Superfusion of the cortical surface with fluorocarbons that contain a high concentration of O_2 attenuates vasodilatation associated with seizures (*Kontos* et al. 1978b). This finding suggests that tissue PO_2 decreases under these conditions.

The potassium ion is also a possible factor coupling cerebral metabolism and blood flow. Electrical or physiological activation results in an increase in ECF [K^+] (*Prince* et al. 1973; *Galvan* et al. 1979; *Singer* and *Lux* 1973, 1975; *Futamachi* et al. 1974; *Urbanics* et al. 1978; *Heinemann* et al. 1977). In addition, K^+ is a potent dilator of pial arteries in the range 0–10 mmol (*Kuschinsky* et al. 1972), and increased CSF [K^+] increases CBF (*Cameron* and *Caronna* 1976). During bicuculline-induced seizures, ECF [K^+] increases from the onset of the response (*Astrup* et al. 1979). Extracellular fluid [K^+] increased from 3.4 to approximately 7 mmol/l at a time when CBF increased four- to fivefold (*Astrup* et al. 1979). In addition, during electrical stimulation of the cortex, ECF [K^+] increases approximately 2 mEq/l as CBF increases (*Heuser* 1979), although this increase is transient (*Urbanics* et al. 1978). However, it is unclear whether ECF [K^+] levels stay elevated for more than 1 min during seizures, even though CBF remains elevated. Glial cells, with potent K^+ uptake systems, may be able to restore ECF [K^+] toward normal after the initial increase (*Orkland* et al. 1966). Thus, although increases in ECF [K^+] may contribute to initial hyperemia during electrical stimulation of the cortex and seizures, it is unclear whether [K^+] contributes in later stages.

Adenosine has attracted a great deal of attention as a possible mediator of dilatation in several regional circulations. It has been suggested that adenosine links metabolism and blood flow in the heart (*Berne* 1980). In the brain, adenosine concentration corresponds temporarily to increases in CBF during seizures. Adenosine concentration rises within the first few seconds of seizures (*Winn* et al. 1980b). Similarly, brain adenosine concentrations increased during electrically induced convulsions (*Schultz* and *Lowenstein* 1978). Low voltage stimulation in vivo (*Rubio* et al. 1975) and in vitro (*Pull* and *McIlwain* 1972) increased cerebral tissue adenosine levels. Adenosine may also contribute to maintenance of resting CBF. Intra-arterial infusion of dipyridamole, which inhibits uptake of adenosine and leads to accumulation of adenosine in ECF, increases CBF during control conditions (*Heistad* et al. 1981). Currently available evidence suggests that adenosine contributes to the coupling of blood flow and metabolism in the brain.

Other factors also may link cerebral metabolism and blood flow. ECF [Ca^{2+}] decreases at the onset of seizures or electrical stimulation of the cortex, when CBF increases (*Heuser* 1979). In addition, ECF [Ca^{2+}] decreases in the sensory cortex during tactile stimulation of skin afferents (*Heinemann* et al. 1977). ECF [Ca^{2+}] is an important determinant of cerebrovascular tone, as reduced ECF [Ca^{2+}] produces vasodilatation (*Betz* and *Csornai* 1978). Extracellular osmolality could increase during augmented cerebral metabolism, as occurs in other tissues (*Jonsson* 1970), and dilate cerebral arteries (*Wahl* et al. 1973).

6 Autoregulation

6.1 Pressure/Flow Relationship

A characteristic feature of the cerebral circulation is that CBF is maintained relatively constant over a wide range of perfusion pressure. Perfusion pressure in the cerebral circulation is defined as arterial pressure minus either cerebral venous pressure or CSF pressure, whichever of the latter two is greater. Under most physiological conditions, cerebral venous and CSF pressures are relatively low, and are often assumed to be negligible.

Under steady state conditions, in anesthetized as well as awake adult animals, autoregulatory mechanisms prevent CBF from changing appreciably over the arterial pressure range of approximately 60–150 mmHg (*Rapela* and *Green* 1964; *Harper* 1966). However, when arterial pressure falls below 50 mmHg, perfusion pressure is not sufficient to maintain CBF (*Harper* 1966; *Marcus* and *Heistad* 1979), even though cerebral resistance vessels have not dilated maximally (*Kontos* et al. 1978a). At the other extreme, when arterial pressure is abruptly increased above 160 mmHg, the autoregulatory capacity of cerebral vessels is exceeded, so that CBF increases and the blood-brain barrier is disrupted (*Bill* and *Linder* 1976; *MacKenzie* et al. 1976; *Heistad* and *Marcus* 1979; *Strandgaard* et al. 1974).

There are important regional differences in the responsiveness of cerebral vessels to changes in arterial pressure during severe hypotension (arterial pressure <50 mmHg) or severe hypertension (arterial pressure >160 mmHg). During severe hypotension, blood flow to the cerebral cortex and medulla are maintained better than to other areas of the brain (*Mueller* et al. 1977; *Marcus* and *Heistad* 1979). During severe hypertension, blood flow to the brainstem increases less than in other areas of the brain (*Baumbach* and *Heistad* 1983). The mechanism responsible for regional differences in autoregulatory capacity is unclear, but it appears that blood flow to more crucial areas of the brain is maintained relatively normal even during extremes of arterial pressure.

CBF also is maintained relatively constant over a wide range of arterial pressure in newborns. In newborn dogs, aged 1–7 days, CBF did not change significantly over the range 27–97 mmHg (*Hernandez* et al. 1980). The maintenance of CBF at control levels even at 40 mmHg and below, in contrast to the situation in adults in which CBF falls appreciably at these pressures, may reflect adaptations of the cerebral circulation of fetuses and newborns to lower resting arterial pressure.

Under certain pathological conditions, such as hydrocephalus and head trauma, CSF or cerebral venous pressures can increase and become important determinants of cerebral perfusion pressure. In general, it appears that CBF is maintained until perfusion pressure falls below approximately 50–60 mmHg. In anesthetized rabbits, CBF began to fall below control levels when cerebral perfusion pressure was decreased to 52 mmHg by elevation of intracranial CSF pressure (*Sadoshima* et al. 1981b). Further elevations of intracranial pressure progressively reduced CBF. Under these conditions, blood flow to the medulla was better preserved than blood flow to other areas of the brain. In addition, *Wagner* (1982) reported that CBF in anesthetized dogs was maintained until perfusion pressure was reduced to 50 mmHg, regardless of whether cerebral venous or CSF pressure was increased.

The temporal course of autoregulatory responses has not been studied extensively but appears to be very rapid. *Nornes* et al. (1977) measured blood flow through the middle cerebral artery in several patients undergoing surgery and provided preliminary evidence that cerebral arteries in man begin to constrict within 10 s after a sudden increase in arterial pressure. Similarly, *Symon* et al. (1973) measured outflow through Labbe's vein and found that blood flow returned toward control levels within 10–90 s of the onset of hypertension. *Kontos* et al. (1978a) found that steady state decreases in cerebral arterial caliber occurred in less than 60 s after attainment of stable hypertension. These authors also found that vasodilatation of pial arteries followed abrupt drops in arterial pressure with a delay time of only a few seconds. Recently, we have applied two different methods to the cerebral circulation in order to characterize responses to abrupt, moderate increases in arterial pressure in anesthetized dogs (*Busija* et al. 1980a). First, blood flow velocity in a branch of the middle cerebral artery was measured continuously using a pulsed-Doppler flow probe. When arterial pressure was increased from approximately 80 to 130 mmHg by aortic constriction, blood flow velocity increased by 50% at the onset of hypertension, but returned almost to baseline within 10 s, despite sustained hypertension. Second, microspheres were used to measure CBF during similar increases in arterial pressure. At the onset of hypertension, CBF increased by 50%, but within 10 s CBF had returned halfway to baseline. Thus, cerebral vascular responses to elevations of arterial pressure are very rapid.

In apparent contrast to other circulations, large cerebral arteries, including the internal carotid artery, circle of Willis, and large pial arteries, as well as smaller arteries, are important resistance vessels (Fig. 2). Both large and small arteries contribute to autoregulatory adjustments in resistance

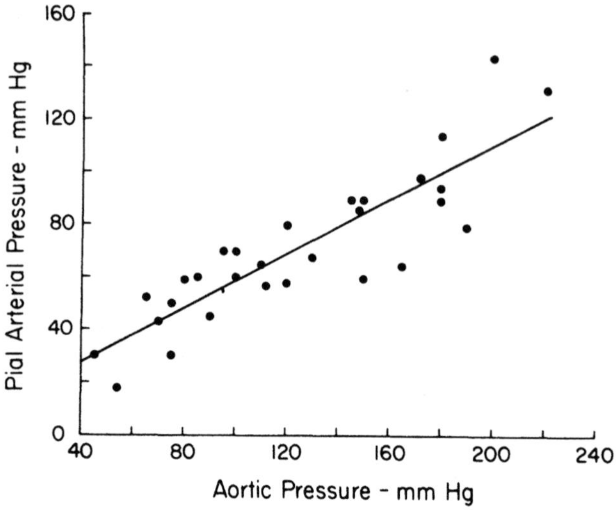

Fig. 2. Relationship between intravascular pressure in pial arteries 150–225 μm in diameter and corresponding aortic pressure during steady state changes in the latter. At a given aortic pressure, pial artery pressure is attenuated substantially. (After *Kontos* et al. 1978a)

to alterations in arterial blood pressure, but the relative importance of each component is dependent on the level of arterial blood pressure. Small arteries predominate in importance when arterial pressure is between approximately 70 and 100 mmHg (*Kontos* et al. 1978a; *Baumbach* and *Heistad* 1983), while resistance changes of larger arteries predominate when arterial pressure is 110–160 mmHg (*Kontos* et al. 1978a; *Baumbach* and *Heistad* 1983). When arterial pressure is elevated beyond 160–200 mmHg, the large pial arteries remain constricted but smaller arteries begin to dilate (*MacKenzie* et al. 1976b; *Kontos* et al. 1978a). Dilatation of the smaller arteries represents loss of autoregulation and is accompanied by disruption of the blood-brain barrier. Thus, the "sausage-string" appearance of cerebral arteries during severe hypertension represents areas of autoregulatory constriction and passive dilatation, not spasm of arteries (*Farrar* et al. 1976).

6.2 Factors Affecting Autoregulation

Several factors can affect the normal pressure-flow relationship of the cerebral circulation. These factors include: 1) cerebral vasodilatation; 2) chronic hypertension; 3) trauma; 4) atherosclerosis; 5) sympathetic nervous activity (see Sects. 7.1.2 and 7.1.3).

6.2.1 Cerebral Vascular Dilatation

During severe hypercapnia and hypoxia, cerebral arteries are dilated and there is a pronounced increase in CBF. Under these conditions, cerebral autoregulation is inhibited, as CBF passively follows arterial pressure (*Häggendal* and *Johansson* 1965; *Harper* 1966; *Kogure* et al. 1970a; *Busija* 1984a). What is the mechanism that is responsible for this inhibition of autoregulation? Because cerebral vasodilatation is already substantial, additional capacity for dilatation to compensate for a fall in arterial pressure is small. During increases in pressure, acidosis and metabolic factors inhibit autoregulatory vasoconstriction, so that increases in arterial pressure produce passive increases in CBF.

The resting tone of cerebral arteries also is a factor in determining resistance to disruption of the blood-brain barrier. In anesthetized rats, cerebral vasodilatation by hypercapnia or papavarine has been reported to enhance disruption of the blood-brain barrier during increases in arterial pressure (*Hardebo* 1981).

6.2.2 Chronic Hypertension

During sustained, prolonged hypertension, cerebral vessels become thickened due to structural changes which include hypertrophy of smooth muscle cells (*Johansson* and *Nordborg* 1978; *Hart* et al. 1980). These changes represent an adaptation to hypertension, so that increases in wall stress, which is inversely proportional to the thickness of the arterial wall (tension = pressure × radius ÷ wall thickness), are attenuated. As a result of these structural changes, one would expect constrictor responses to be augmented and maximal dilatation to be impaired (*Folkow* and *Neil* 1971). We found that during bicuculline-induced seizures, resistance was twofold higher in spontaneously hypertensive rats than in normotensive rats (*Sadoshima* et al. 1983).

Although resting CBF is not different in hypertensive and normotensive man and animals (*Strandgaard* et al. 1973, 1975; *Strandgaard* 1976; *Jones* et al. 1976; *Sadoshima* et al. 1983), the pressure-flow relationship for the cerebral circulation is shifted toward the right. The result is that when arterial pressure is lowered, CBF in hypertensive animals and man begins to fall from control values at higher arterial pressures than in normotensives (*Strandgaard* 1976; *Jones* et al. 1976; *Barry* et al. 1982). Furthermore, CBF in hypertensive animals begins to increase from control values at higher arterial pressures than in normotensives when arterial pressure is elevated (*Strandgaard* et al. 1973, 1975). There are data which suggest that this effect of chronic hypertension on the pressure-flow relationship is reversible. In previously hypertensive patients and rats, when arterial pressure was lowered pharmacologically, there was reversion toward a normal pressure-flow relationship (*Strandgaard* 1976; *Barry* et al. 1983).

Structural changes in cerebral vessels produced by chronic hypertension also protect against disruption of the blood-brain barrier. During similar elevations in arterial pressure, disruption of the blood-brain barrier to albumin was reduced by one-half in spontaneously hypertensive rats, compared with normotensive rats (*Mueller* and *Heistad* 1980). This effect is not observed in all models of hypertension. The blood-brain barrier is more susceptible to disruption in renal hypertensive rats than in normotensive rats (*Johansson* and *Linder* 1980; *Mueller* and *Luft* 1982).

6.2.3 Trauma

Trauma to cerebral arteries may reduce normal responsiveness to alterations in arterial pressure. For example, excessive surgical manipulation of cerebral vessels eliminates autoregulation (*Sagawa* and *Guyton* 1961). In addition, concussive head trauma in animals and man leads to impairment of cerebral vascular responses to alteration in arterial pressure (*Langfitt* et al. 1966; *Wei* et al. 1980a). The mechanism for this altered responsiveness appears to be the increase in arterial pressure that accompanies head trauma (*Wei* et al. 1980a). Transient but profound hypertension after head trauma causes damage to cerebral arteries, such that resting oxygen consumption of vessels is reduced, the endothelium is pitted by lesions, and the vessels exhibit sustained vasodilatation. When arterial pressure subsequently was lowered to normotensive levels, pial arteries had reduced or absent ability to undergo normal autoregulatory vasodilatation.

6.2.4 Atherosclerosis

Atherosclerosis affects primarily large cerebral arteries (*Solberg* and *Egger* 1971; *Stehbens* 1972), and since these arteries are important resistance vessels, one would expect CBF responses to be affected by atherosclerosis. However, in atherosclerotic monkeys, resting CBF does not differ from that in normal monkeys (*Heistad* et al. 1980b). Because large cerebral arteries are partially occluded by atherosclerotic lesions, it seems likely that small, distal cerebral arteries must dilate to preserve resting CBF. When arterial pressure is decreased, autoregulation is intact, although cerebral vascular dilatation during hypercapnia is hindered (*Heistad* et al. 1980b). Resorption of cerebral atherosclerotic lesions is accompanied by a marked improvement in cerebral vasodilator responses (*Armstrong* et al. 1983).

6.3 Mechanisms of Autoregulation

Considerable controversy exists concerning the mechanisms underlying cerebral autoregulation. Possible mechanisms include neurogenic, myogenic, and metabolic factors.

Although early studies indicated that activation of arterial baroreceptors leads to cerebral vasoconstriction via reflex pathways in baboons (*Ponte* and *Purves* 1974; *James* and *MacDonell* 1975), more recent studies in monkeys and dogs have suggested that arterial baroreceptors are not involved in this response (*Heistad* and *Marcus* 1976; *Rapela* et al. 1976). In addition, section of sympathetic nerves supplying cerebral vessels (*Mueller* et al. 1977; *Busija* 1983b) or stimulation of nerves putatively supplying dilator fibers to cerebral vessels (*Linder* 1981) does not alter cerebrovascular responses to alterations in arterial pressure within the physiological range. Thus, it seems unlikely that nerves are necessary for autoregulatory responses of cerebral vessels. Nonetheless, as described in the section on neural control, sympathetic nerves modify the pressure-flow relationship during severe hypertension and perhaps hypotension (see Sects. 7.1.2 and 8.1.3).

It has been suggested that myogenic factors are involved in autoregulatory responses, as changes in intravascular pressure constitute the stimulus to contraction or relaxation of smooth muscle in cerebral arteries. There is evidence that myogenic factors affect resistance in other vascular beds. For example, stretch or relaxation of rat portal vein in vitro elicits coordinated electrical and mechanical activity; the rate of change of stretch or relaxation is an important component of the response (*Johansson* and *Mellander* 1975). In the cerebral circulation, *Toda* et al. (1978) found that stretching a cerebral artery in the presence of serotonin in vitro caused contraction. In addition, *Vindall* and *Simeone* (1982) found that increased intravascular pressure causes contraction of anterior and middle cerebral arteries in vitro. Thus, myogenic factors may play a role in autoregulatory responses.

Metabolic factors also have been proposed as important mediators of autoregulation. Changes in perfusion pressure may alter local concentrations of vasoactive substances which may affect vascular tone. Metabolic factors currently thought to play an important role in cerebral autoregulation include ECF [H^+], [K^+], and adenosine concentration. All three of these substances are potent cerebral vasodilators (*Kuschinsky* et al. 1972; *Heistad* et al. 1981). However, *Wahl* and *Kuschinsky* (1979) measured perivascular [H^+] and [K^+] near pial vessels when arterial pressure was varied over the range 60–200 mmHg, and found that the concentrations of these substances did not change even though pial arteries constricted during increases in arterial pressure and dilated when arterial pressure was lowered. This study supports earlier reports by *Eklöf* et al. (1972) and *Siesjö* and *Zwentnow* (1970). In contrast, brain adenosine concentration has been found to double when arterial blood pressure is decreased from 135 to 72 mmHg, and to increase sixfold when arterial blood pressure is decreased from 135 to 45 mmHg (*Winn* et al. 1980a). Changes of this magnitude in adenosine concentration are consistent with adenosine playing an important role in autoregulatory responses. However, it is not clear whether brain levels of adenosine accurately reflect the concentration of adenosine near cerebrovascular smooth muscle.

Another approach that has been used to differentiate between myogenic and metabolic factors is to increase cerebral venous pressure and observe the responses of pial arteries or CBF. This procedure puts the two factors into opposition. It tends to reduce perfusion pressure and increases concentrations of vasoactive metabolites on the one hand, and raises intravascular pressure and increases the myogenic stimulus on the other (*Johnson* and *Henrich* 1975). When *Wei* and *Kontos* (1982) increased cerebral venous pressure with vena caval occlusion, pial arteries dilated, which is consistent with a maintenance of CBF. In similar studies, CBF did not change when cerebral venous pressure was elevated in dogs (*Wagner* and *Traystman* 1983). These findings are consistent with metabolic factors playing an important role in autoregulatory response. Thus, myogenic mechanisms have been demonstrated in vitro, but studies in vivo suggest that metabolic factors predominate.

7 Neural Effects

The question of the contribution of nerves to the regulation of the cerebral circulation has been very controversial (*Heistad* and *Marcus* 1978; *Purves* 1978). Much of this controversy has been due to the use of different and sometimes inappropriate methods for measuring CBF, uncertainty regarding origins of nonsympathetic nerves, and species differences (*Heistad* and *Marcus* 1978; *Busija* and *Heistad* 1981; *Busija* et al. 1982b). However, methods are sufficiently advanced to provide a basic overall understanding of the role played by nerves in the regulation of the cerebral circulation.

Cerebral vessels are richly supplied by at least three types of autonomic nerves: sympathetic (*Nielsen* and *Owman* 1967); cholinergic (*Lindvall* et al. 1975; *Florence* and *Bevan* 1979); and peptidergic (*Larsson* et al. 1976; *Chan-Palay* 1977; *Edvinsson* et al. 1981). There is preliminary evidence that cerebral vessels receive other types of peripheral innervation (*Chan-Palay* 1977; *Peerless* and *Kendall* 1975). In addition, it has been suggested that neural systems, possibly originating in the locus coeruleus and other brain areas, provide innervation to cerebral vessels via direct central nervous system pathways (*Hartman* et al. 1972; *Nakai* et al. 1983).

This section will review studies concerning the functional importance of innervation of cerebral vessels. Because the origins, distribution, and effects of sympathetic innervation are well known, in contrast to cholinergic, peptidergic, and central innervations, this section will emphasize the functional significance of sympathetic nerves.

7.1 Sympathetic Nerves

Cerebral vessels are supplied by sympathetic nerves that originate predominantly from the superior cervical ganglia (*Nielsen* and *Owman* 1967; *Edvinsson* et al. 1972b; *Mueller* et al. 1977). Postganglionic fibers originating from these ganglia form a well-developed plexus in cerebral arteries and arterioles (*Nielsen* and *Owman* 1967; *Peerless* and *Yasargil* 1971; *Edvinsson* et al. 1976a; *Purdy* and *Bevan* 1977). Cerebral veins are also innervated but the density is lower than in arteries (*Nielsen* and *Owman* 1967). Innervation is primarily ipsilateral in the middle cerebral artery (*Nielsen* and *Owman* 1967), but there is overlap in vessels in basal and medial areas of the brain, such as the circle of Willis, anterior cerebral artery, and basilar artery (*Nielsen* and *Owman* 1967; *Peerless* and *Yasargil* 1971). In some species, cerebral vessels in caudal areas of the brain are supplied to a limited extent by fibers originating from sympathetic ganglia other than the superior cervical ganglia (*Edvinsson* 1975).

In general, density of innervation is similar to mesenteric and femoral vascular beds (*Rosenblum* 1976a,b) and is very heterogeneous. Innervation of arteries arising from the internal carotid system is denser than vessels of the vertebral system. Approximately 60%–90% of the arterioles in the medial geniculate body, parietal and temporal cortices, caudate nucleus, inferior colliculus, thalamus, and hypothalamus are innervated, in contrast to 10%–30% of arterioles in the medulla, occipital cortex, and cerebellum (*Edvinsson* and *Owman* 1977). Variation in the density of innervation is suggestive of preferential sympathetic effects in certain areas of the brain.

7.1.1 In Vitro Studies

Studies of isolated cerebral arteries have provided valuable information that cannot be obtained by other methods concerning neuromuscular mechanisms. It appears that α-adrenergic receptors in cerebral arteries have characteristics different from those found in other regional circulations. Thus, α-receptors of cerebral arteries are relatively insensitive to norepinephrine (*Bevan* and *Bevan* 1977; *Duckles* and *Bevan* 1976), exhibit little stereospecifictiy between *d*- and *l*-norepinephrine (*Duckles* and *Bevan* 1976), have relatively high dissociation constants for norepinephrine and phentolamine (*Edvinsson* and *Owman* 1974), and relative potencies of various agonists are unusual (*Edvinsson* and *Owman* 1974). The relative insensitivity of the basilar artery may be due to unusual receptor-contraction coupling mechanism. In contrast to other peripheral arteries, cerebral arteries apparently rely almost exclusively on extracellular calcium for norepinephrine-induced contraction (*McCalden* and *Bevan* 1981; *Towart* 1981; *Towart* and *Kazda* 1982). In contrast, release of intracellular calcium may contribute to intrinsic tone (*Bevan* 1983).

Several recent studies have provided new data concerning electrogenic mechanisms in cerebral arteries. The membrane potential or smooth muscle of the basilar artery was found to be between –52 mV (*Fujiwara* et al. 1982) and –62 mV (*Harder* et al. 1981) and that of the middle cerebral artery was found to be –52 mV (*Harder* et al. 1981; *Fujiwara* et al. 1982; *Suzuki* and *Fujiwara* 1982). The membrane potential for cerebral arteries is higher than that for mesenteric arteries, due to a higher K^+ conductance and a higher electrogenic Na-K pump during resting conditions (*Harder* 1980). *Harder* et al. (1981) found good correlation between membrane potential and tension generated with all but very high doses of norepinephrine in cat cerebral arteries. Norepinephrine depolarized and contracted the basilar artery but hyperpolarized and relaxed the middle cerebral artery. The former effect was blocked by phentolamine and the latter by propranolol. In contrast, serotonin depolarized and contracted both vessels. *Lusamvuku* et al. (1979) also found spike frequency in the middle cerebral artery of the rabbit correlated with tension generated by norepinephrine. Application of tetraethylammonium chloride (TEA), which suppresses K^+ permeability, does not change membrane potential in the basilar artery but evokes spike potentials upon application of norepinephrine (*Harder* et al. 1981). For similar effects in the middle cerebral artery, TEA plus combined norepinephrine and propranolol are needed for generation of spike potential (*Harder* et al. 1981). Similarly, transmural nerve stimulation in the dog basilar artery evokes an initial, fast excitatory depolarization, followed by a slower hyperpolarization (*Fujiwara* et al. 1982; *Suzuki* and *Fujiwara* 1982). The depolarization is abolished by treatment with 6-hydroxydopamine but the hyperpolarization is unaffected by atropine or purinergic 1 and 2 receptor blockade.

Transmural electrical stimulation of cerebral arteries activates intramural sympathetic nerves and produces constriction in pial and basilar arteries of rabbits, monkeys, sheep, and dogs (*Lee* et al. 1976; *Duckles* et al. 1977; *Duckles* and *Bevan* 1979). In contrast, in cats and pigs, transmural electrical stimulation results primarily in relaxation (see Sect. 7.2) (*Lee* 1982; *Winquist* et al. 1982). The vasoconstrictor response to transmural stimulation is abolished by sympathetic denervation or guanethidine (*Lee* et al. 1976; *Duckles* 1979), but is not attenuated by α-adrenergic blockade, except perhaps in the initial phase, and in fact may be augmented in rabbits (*Lee* et al. 1976, 1980). However, vasoconstriction is blocked by phentolamine in sheep (*Duckles* 1979). In addition, phentolamine augments vasoconstriction and norepinephrine overflow during transmural stimulation (*Duckles* and *Rapoport* 1979; *Muramatsu* et al. 1977). The response to transmural stimulation is similar to the response to exogenous norepinephrine in that maximal contraction is much less than that induced by histamine or serotonin (*Bevan* et al. 1975; *Duckles* and *Bevan* 1979).

In general, other arteries are much more responsive to exogenous norepinephrine or transmural stimulation than cerebral arteries (*Bevan* and *Bevan* 1977). *Bevan* (1979) has suggested that differences in sensitivity between cerebral and noncerebral arteries are related to different embryonic origins. Extracranial segments of carotid and vertebral arteries are nearly 100 times more responsive to norepinephrine than intracranial segments of the same vessels. Intra- and extracranial arteries arise from different primordial mesodermal cells, and the site of fusion corresponds to the abrupt change in responsiveness.

7.1.2 In Vivo Studies

Electrical stimulation of sympathetic nerves that supply cerebral vessels has been used to simulate reflex activation and presumably produces maximal effects on CBF. Because sympathetic fibers also supply extracerebral tissue such as cranial muscle, it is possible to document the efficacy of stimulation and compare the effects of nerves on cerebral and extracerebral tissues simultaneously. During normotension, supramaximal sympathetic stimulation has little or no effect on CBF in cats and dogs (*Alm* and *Bill* 1973; *Heistad* et al. 1977, 1978b) but decreases CBF 20%–30% in monkeys and rabbits (*Lacombe* et al. 1977; *Heistad* et al. 1978b; *Busija* 1984b). In all species, sympathetic stimulation reduces cranial muscle blood flow almost to zero. Responses in rabbits and monkeys occur primarily in cortical gray matter and the caudate nucleus. In monkeys , unilateral sympathetic stimulation decreases blood flow to cerebral gray and white matter and the caudate by 15%–25% initially. But during continued stimulation, blood flow returned to control over 5 min (*Marcus* et al. 1982). In contrast, sympathetic stimulation reduces CBF in rabbits by up to 30% and only caudate blood flow "escapes" toward control (*Sercombe* et al. 1979; *Marcus* et al. 1982). Cranial muscle blood flow does not "escape" during several minutes of stimulation. During acute severe hypertension in cats, cerebral vessels do not escape from sympathetic constrictor effects (*Marcus* et al. 1982).

Recent preliminary studies in rabbits suggest that bilateral activation of sympathetic nerves during normocapnia reduces CBF more than unilateral stimulation (*Busija* 1983b). Bilateral stimulation decreased blood flow to cortical gray matter by 18%–23% at 4 and 8 Hz, but unilateral stimulation did not affect blood flow.

There have been two areas of apparent conflict concerning sympathetic effects on cerebral vessels, based on studies using microspheres, on the one hand, and the cranial window or venous outflow methods, on the other. Using the cranial window method, several investigators (*Kuschinsky* and *Wahl* 1975; *Wei* et al. 1975; *Auer* et al. 1981) found that pial arterial diameter decreased 7%–13% during sympathetic stimulation in cats. In contrast, CBF measured with microspheres under these conditions in cats does not decrease (*Alm* and *Bill* 1973; *Boisvert* et al. 1977; *Heistad* et al. 1978b). If a 7%–13% reduction occurred in all cerebral resistance vessels, one would expect CBF to decrease 20%–25%. To resolve this controversy, we studied the effects of sympathetic nerves on cerebral arteries in cats using a new method that permits continuous measurement of diameter and continuous measurement of flow velocity in pial arteries (*Busija* et al. 1981). During sympathetic stimulation, the diameter of large arteries (>150 μm) decreased 6% ± 1%, but calculated CBF did not change as blood flow velocity increased 19% ± 5% (*Busija* et al. 1982a). Apparently, downstream arteries dilate in response to a fall in intraluminal pressure when larger arteries constrict, confirming the hypothesis of *Harper* et al. (1972). The question of the effects of sympathetic nerves on CBF in dogs has been particularly controversial. *D'Alecy* and colleagues (*D'Alecy* and *Feigl* 1972; *D'Alecy* et al. 1979), using a simplified version of the venous outflow method, reported that sympathetic stimulation markedly decreased CBF (up to 80%) in anesthetized dogs, while another group using a different venous outflow method (*Traystman* and *Rapela* 1975), and others using microspheres (*Heistad* et al. 1978b), found that CBF did not change during sympathetic stimulation during normotension in dogs. It seems likely that this discrepancy is due to a methodological limitation of the simplified version of the venous outflow method of *D'Alecy* and colleagues. A recent study has shown that the retroglenoid vein, a continuation of the temporal sinus, which carries all the venous effluent in the simplified venous outflow preparation, is heavily innervated by sympathetic fibers from the stellate ganglia and is very responsive to electrical stimulation (*Pearce* et al. 1982). During stellate stimulation in dogs, the retroglenoid vein may constrict, raise pressure in intracerebral veins, and divert cerebral venous outflow through other veins, thus leading to an underestimation of CBF (*Pearce* et al. 1982). A recent study by *Wagner* and *Traystman* (1983) suggests that an increase in cerebral venous outflow pressure opens intracranial venous anastomoses and diverts outflow from the confluence of sinuses to other veins draining the brain.

In contrast to effects during normotension, sympathetic nerves have been found consistently to have important effects on CBF when arterial pressure is elevated, even in species that normally show few sympathetic effects. When arterial pressure is increased abruptly within the physiological range of pressure by aortic constriction in anesthetized cats, CBF increases 50% transiently before autoregulatory adjustments return CBF to control (*Busija* et al. 1980a). When sympathetic nerves are stimulated, this transient increase in CBF is blunted severely. Sympathetic effects are also pronounced during maintained, severe elevations of arterial pressure. When mean arterial pressure is increased to 170–220 mmHg, the autoregulatory capacity of cerebral arteries is exceeded and CBF increases severalfold. *Bill* and *Linder* (1976) first demonstrated that sympathetic stimulation in cats greatly attenuates this increase in CBF during hypertension. This finding has been confirmed in several laboratories (*Edvinsson* et al. 1976b; *Heistad* et al. 1978b; *MacKenzie* et al. 1979). *MacKenzie* et al. (1979) found that bilateral sympathetic stimulation attenuated the increase in CBF more than unilateral stimulation.

During severe hypertension, the blood-brain barrier, which normally minimizes entry of substances such as albumin from blood to ECF, is disrupted (*Bill* and *Linder* 1976; *Heistad* and *Marcus* 1979). Under these circumstances, electrical and physiological activation of sympathetic nerves not only attenuates increases in CBF, but also reduces extravasation of albumin into the brain (*Bill* and *Linder* 1976; *Heistad* and *Marcus* 1979; *Gross* et al. 1979). The physiological value of this "protection" is clear. Pronounced increases in arterial pressure, due primarily to activation of the sympathetic nervous system, occur frequently in man and animals during stress (*Bevan* et al. 1969; *Langer* et al. 1979; *Diepstra* et al. 1980). Under these conditions, arterial pressure is able to rise because of suppression of the baroreflex. It is likely that simultaneous activation of sympathetic nerves supplying cerebral vessels could limit passive distension of cerebral arteries and cerebral hyperemia and prevent damage to cerebral vessels.

Several mechanisms may be involved in potentiation of sympathetic effects during hypertension. First, stretching of cerebral arteries during hypertension may increase responsiveness to constrictor stimuli (*Toda* et al. 1978). Second, increased resistance of large arteries, due to sympathetic stimulation, may attenuate increases in intraluminal pressure in smaller, downstream arteries, and thus prevent passive dilatation and vessel damage. Third, sympathetic stimulation may reduce compliance of cerebral arteries and attenuate passive dilatation.

Other conditions also potentiate sympathetic effects of CBF. During hypoxia in lambs, when baseline CBF is elevated, unilateral sympathetic stimulation decreases CBF by 25% (*Wagerle* et al. 1983). During systemic hypercapnia, CBF also increases three- to fourfold. Under these conditions in anesthetized cats, unilateral stimulation increases CVR by 15% ± 4%, and bilateral stimulation increases CVR by 60% ± 16% in cats (*Busija* and *Heistad* 1984). During hypercapnia, sympathetic stimulation constricts large cerebral arteries, and because smaller arteries are dilated already, CBF falls. There are at least two possible explanations for the finding that bilateral effects are greater than unilateral effects. First, differences in unilateral and bilateral effects may be due to differential responsiveness of two sets of cerebral resistance vessels – arteries proximal to and including the circle of Willis on the one hand and, on the other, relatively large pial arteries distal to these vessels. Although presumably both proximal and distal arteries on the ipsilateral side constrict when nerves on one side are stimulated, contralateral blood flow may increase in proximal arteries, so that reductions in ipsilateral hemispheric blood flow are small. During bilateral stimulation, both sets of resistance vessels are simultaneously constricted, so that a compensatory increase in blood flow cannot occur. An alternative hypothesis is that simultaneous activation of sympathetic nerves arising from the contralateral side may potentiate cerebral vasoconstriction during activation of ipsilateral sympathetic nerves, i.e., the summation of stimuli may produce a shift to a steeper segment of the stimulus-response relationship.

There are a number of neural and vasoactive substances that exist contiguously and may interact in affecting cerebral vessels. Cholinergic mechanisms may attenuate sympathetic vasoconstriction in the cerebral circulation (see Sect. 7.2.2). In addition, *Gross* et al. (1983) have found that histamine attenuates the effects of sympathetic nerves on pial arteries and veins. In contrast, inhibition of endogenous prostaglandin synthesis by indomethacin did not modify sympathetic vasoconstriction in the cerebral circulation (*Busija* 1984b).

The effects of sympathetic nerves on cerebral veins have been examined recently. *Auer* et al. (1981) found that pial veins constrict 12%–20% during nerve stimulation, and this effect is greater on a diameter basis for veins than for pial arteries. An important limitation of studies of pial veins is that it is difficult to interpret changes in venous diameter unless pressure in the vein is known. For example, reduction in microvascular pressure by sympathetic stimulation (*Baumbach* and *Heistad* 1983) may reduce venous pressure, so that decreases in venous diameter may be a reflection of passive "collapse" instead of active constriction. Nevertheless, the capacity of sympathetic nerves to regulate the tone of cerebral capacitance vessels provides further evidence that sympathetic nerves may play an important role in the regulation of cerebral blood volume (*Edvinsson* et al. 1971).

7.1.3 Reflex Activation of Sympathetic Nerves

To demonstrate the importance of neural control, it is necessary to show that sympathetic nerves can alter CBF during physiological conditions. Section of sympathetic nerves does not increase CBF in anesthetized dogs (*Mueller* et al. 1977; *Busija* et al. 1980a), cats (*Heistad* et al. 1978b; *Busija* et al. 1980a), monkeys (*Heistad* et al. 1978b), and rabbits (*Sadoshima* et al. 1981b), or in awake dogs (*Marcus* and *Heistad* 1979), cats (*Heistad* et al. 1980c), and rabbits (*Busija* 1984a). In addition, topical application of phentolamine does not alter the diameter of pial arteries in anesthetized cats (*Kuschinsky* and *Wahl* 1975). Thus, there is apparently little or no resting sympathetic tone to cerebral vessels.

Animals have been exposed to several physiological stresses to examine reflex mechanisms. During systemic hypotension, sympathetic nerves have no effect on CBF in anesthetized cats (*Gross* et al. 1979) and dogs (*Mueller* et al. 1977), or in awake cats (*Heistad* et al. 1980b). However, sympathetic nerves reduce CBF modestly during hypotension when arterial pressure is between 40–60 mmHg in anesthetized monkeys (*Fitch* et al. 1975) and awake dogs (*Marcus* and *Heistad* 1979), but not in awake rabbits (*Busija* 1984a).

Unloading of baroreceptors by sino-aortic nerve deafferentation results in intense sympathetic discharge, a twofold increase in arterial pressure, and a threefold increase in CBF. Reflex activation of sympathetic nerves attenuates the increase in cortical gray blood flow by 24% (*Gross* et al. 1979). In contrast, following lesions of the nucleus tractus solitarii, which result in hypertension and increased CBF, sympathetic nerves do not affect CBF (*Graham* et al. 1982). During similar conditions, such as bicuculline-induced seizures, CBF rises severalfold due to an increase in both arterial pressure and cerebral metabolism. Under these conditions, sympathetic nerves modestly attenuate the increase in CBF (*Mueller* et al. 1979).

Stimulation of carotid baroreceptors and chemoreceptors has been reported to affect CBF via activation of sympathetic nerves. Several investigators have reported that exposure of baroreceptors to increased arterial pressure produces cerebral vasoconstriction in baboons (*Ponte* and *Purves* 1974; *James* and *MacDonell* 1975). In contrast, we (*Heistad* and *Marcus* 1976) and others (*Rapela* et al. 1967) have found that stimulation of baroreceptors does not play a role in the regulation of CBF in dogs and monkeys. It seems likely that the conflicting results are related to differences in experimental procedures and methods used for measuring CBF.

Stimulation of carotid chemoreceptors produces intense vasoconstriction in muscle, kidney, and small bowel, but no effect on CBF in anesthetized dogs or monkeys (*Heistad* et al. 1976). Recently, *Vatner* et al. (1980) examined sympathetic effects on CBF during stimulation of chemoreceptors in awake but paralyzed and artificially ventilated dogs. When sympathetic nerves to cerebral vessels were sectioned bilaterally, CBF increased approximately 40% as arterial pressure increased. However, when sympathetic nerves were intact, CBF did not increase, even though arterial pressure rose. The variation of results appears to result from differences of physiological conditions and effects of chemoreceptor stimulation on arterial pressure. *Heistad* et al. (1976) activated chemoreceptors with hypoxic and hypercapnic blood and measured CBF at a time when arterial pressure was normal. In contrast, *Vatner* et al. (1980) stimulated chemoreceptors with intracarotid nicotine and measured CBF during resultant arterial hypertension. As noted previously, cerebral vasoconstrictor effects of sympathetic nerves are potentiated by acute hypertension (*Heistad* et al. 1978b; *Busija* et al. 1980a).

Systemic hypercapnia and hypoxia increase CBF and activate sympathetic pathways to cerebral vessels. *Wei* et al. (1980a) found that large pial arteries in anesthetized cats dilated during hypercapnia, but that this dilatation was attenuated by intact sympathetic nerves. Recently, we have found that in anesthetized cats and awake rabbits, when CBF increases to greater than 110 ml/min/100 g during hypercapnia, intact sympathetic nerves contribute approximately 20% to cerebral vascular resistance (*Busija* and *Heistad* 1984). Sympathetic nerves also attenuate increases in CBF during hypoxia. In awake rabbits, CBF increases during hypoxia, and intact sympathetic nerves reduce vasodilatation by up to 20% in gray matter (*Busija* 1984a).

Thus, reflex activation of sympathetic nerves during various conditions, including hypertension, hypoxia, hypercapnia, and possible hemorrhagic hypotension, has significant effects on cerebral vessels.

7.1.4 Other Effects of Sympathetic Nerves on Cerebral Vessels

In addition to direct, acute effects on CBF, sympathetic nerves have other, more subtle, influences on cerebral vessels. These influences include trophic effects and effects on CSF production.

Sympathetic nerves appear to play a large role in the normal development and protection of cerebral arteries. In normotensive rabbits, sympathectomy results in reduced weight of large cerebral arteries (*Bevan* et al. 1983). In genetically hypertensive rats, intact sympathetic nerves contribute to the development of cerebrovascular hypertrophy and protection of the blood-brain barrier and protect against stroke during chronic hyper-

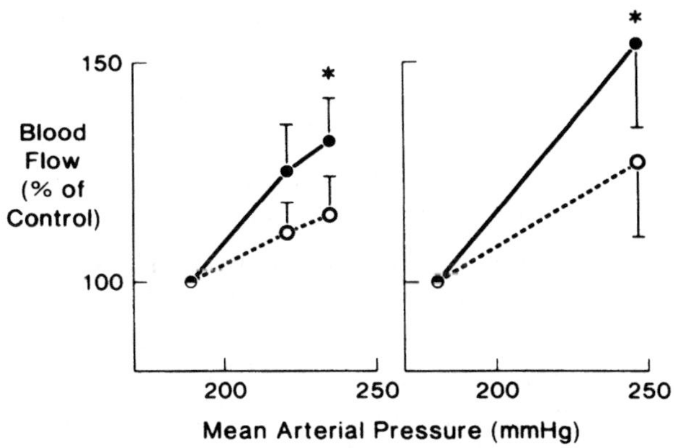

Fig. 3. Effect of elevation of arterial pressure in stroke-prone spontaneously hypertensive rats on blood flow to anterior cerebrum during graded increases in arterial pressure *(left)* and during one-step increase *(right)*. Values are means ± SE. * $P < 0.05$, denervated (●—●) versus innervated (○—○) hemisphere. (After *Sadoshima* et al. 1983)

tension (*Hart* et al. 1980; *Sadoshima* et al. 1981a, 1983; *Mueller* et al. 1982, 1983). Several studies indicate that sympathetic denervation at 1 month, when vessels are growing rapidly, inhibits development of vascular hypertrophy, but denervation at a later age does not affect vascular growth. In stroke-prone spontaneously hypertensive rats (SHR-SP), the wall:lumen ratio in pial arteries (<80 μm in diameter) was less following chronic denervation performed at 1 month than in innervated arteries (*Sadoshima* et al. 1981a, 1983). In contrast, there was no difference in wall:lumen ratio in denervated or intact arteries when denervation was performed 3 months after birth (*Sadoshima* et al. 1983). Similarly, *Mueller* et al. (1982, 1983) found that in spontaneously hypertensive rats, protein transfer across cerebral microvessels during seizures was greater when sympathetic nerves were sectioned at 1 month than in vessels with intact innervation, but there was no difference when denervation was performed at 3 months. *Sadoshima* et al. (1981a, 1983) found that the incidence of stroke was much greater following denervation at 1 month than in animals with intact sympathetic innervation. When denervation was performed at 3 months, there was no difference in incidence of stroke from the side having intact nerves.

Intact sympathetic innervation also appears to be necessary for normal responsiveness of cerebral arteries. During acute elevations of arterial pressure in SHR-SP, CBF increased less in brains with intact sympathetic innervation than in the side denervated at 3–4 weeks after birth (*Sadoshima* et al. 1983: Fig. 3).

The choroid plexus, which is involved in bulk CSF formation, is under autonomic control. Ultrastructural studies indicate that both vessels and epithelial cells of the choroid plexus are innervated by sympathetic nerves (*Edvinsson* et al. 1974, 1975). Physiological studies suggest that sympathetic nerves have important effects on CSF formation. Electrical stimulation of sympathetic nerves reduces CSF formation by up to 30% (*Lindvall* et al. 1978a). In addition, sympathectomy increases CSF formation to the same degree (*Lindvall* et al. 1978a). Thus, it appears that CSF formation is normally under sympathetic inhibitory control. *Lindvall* and *Owman* (1981) have suggested that effects of nerves on CSF formation are exerted by plexus epithelium and blood vessels. Apparently, activation of β-receptors in epithelium inhibits CSF formation directly, and activation of vascular α-receptors reduces choroid plexus blood flow.

7.2 Cholinergic Nerves

Evidence provided by several different approaches indicates that cerebral arteries have a rich cholinergic innervation. First, electron microscopy demonstrates the presence of agranular vesicles, commonly but not exclusively associated with acetylcholine, in nerve terminals on cerebral vessels (*Edvinsson* et al. 1972b). In contrast to sympathetic nerves, these nerves appear to innervate extraparenchymal but not intraparenchymal arteries and not to supply cerebral veins significantly (*Edvinsson* et al. 1972b). There is a close anatomical relationship between these and sympathetic nerves, leading to speculation that one type of fiber can modulate the effects of the other. Second, specific biochemical assays, such as activity of choline acetyltransferase and high-affinity choline uptake, suggest the presence of cholinergic nerves on large cerebral arteries (*Florence* and *Bevan* 1979; *Duckles* 1981b). There is a species difference in the extent of these markers; innervation in cats and rabbits is apparently higher than in dogs (*Florence* and *Bevan* 1979). Third, acetylcholine is present in cerebral arteries, and transmural nerve stimulation of cerebral arteries in vitro results in release of this neurotransmitter (*Duckles* 1981c). Release of acetylcholine under these conditions is inhibited by β-bungarotoxin and hexamethonium (*Duckles* 1981c).

The origin of cholinergic innervation of cerebral arteries is not well defined. Traditionally, the most likely source of this innervation has been thought to be the greater superficial petrosal nerve (GSPN). *Chorobski* and *Penfield* (1932) described the passing of nerve fibers from the GSPN onto the carotid artery. Recently, *Vasquez* and *Purves* (1979), following axonal transport of cobalt chloride, traced only a few fibers from this nerve to cerebral vessels. However, chronic section of the GSPN does not decrease acetylcholinesterase activity or cause degeneration of nerve endings associated with cerebral arteries (*Vasquez* and *Purves* 1979). Nonetheless, the presence of ganglia along the GSPN distal to the section may have prevented axonal transport of tracer and degeneration of postganglionic cholinergic nerves (*Napolitano* et al. 1965). In contrast to sympathetic innervation, the origins of cholinergic innervation may be more diffuse. For example, it has been suggested that cranial nerves III, VII, IX, and X each provide some cerebrovascular innervation (*Pearless* and *Kendall* 1975).

7.2.1 In Vitro Studies

In vitro, acetylcholine relaxes cerebral arteries at low concentrations and constricts them at high concentrations (*Lee* 1982). These effects are blocked by atropine (*Lee* 1982). Following noradrenergic blockade, transmural electrical stimulation causes relaxation (*Bevan* et al. 1982a; *Lee* 1982; *Toda* 1982; *Winquist* et al. 1982). There is a species difference in responsiveness, such that cat and pig arteries relax to a greater extent than arteries of other species (*Lee* 1982; *Winquist* et al. 1982). Presently, there is controversy surrounding the identity of the neurotransmitter involved. *Bevan* et al. (1982a) have reported that atropine substantially attenuates stimulation-induced relaxation. On the other hand, others report that this relaxation is unaffected by β-bungarotoxin (*Duckles* 1981a) or atropine (*Lee* 1980, 1982; *Toda* 1982). In addition, *Lee* et al. (1978) and *Duckles* (1979) report that stimulation-induced relaxation is not potentiated by physostigmine.

The integrity of the endothelium contributes to the response seen in vitro. Following damage to or removal of the endothelium by gentle rubbing of the artery, acetylcholine only causes contraction, while transmural electrical stimulation continues to result in relaxation (*Lee* 1982).

7.2.2 In Vivo Studies

The principal effect of stimulating vascular cholinergic receptors in vivo appears to be dilatation. For example, topical application of acetylcholine causes dilatation of pial arteries, and this response is blocked by atropine (*Kuschinsky* et al. 1974; *Busija* and *Heistad* 1982). Intra-arterial infusion of acetylcholine increases CBF in rabbits (*Heistad* et al. 1980) and rats (*Gross* et al. 1981b).

The effects of GSPN stimulation or section have been controversial. Electrical stimulation has been reported to dilate pial arteries (*Cobb* and *Finesinger* 1932) and increase CBF (*D'Alecy* and *Rose* 1977; *Pinard* et al. 1979). On the other hand, *Busija* and *Heistad* (1981) found that unilateral GSPN stimulation did not increase CBF during normocapnia or hypocapnia in anesthetized cats, although lacrimal gland blood flow increased profoundly. Similar results have been reported by *Scremin* et al. (1983) for cranial nerves III, VII, IX, and X. In addition, *Linder* (1981) and *Scremin* et al. (1983) reported that GSPN stimulation does not affect CBF during hypotension.

Section of the GSPN does not alter resting CBF (*Busija* and *Heistad* 1981). Nerve section has been reported to attenuate markedly the increase in CBF during hypercapnia and hypoxia (*James* et al. 1969; *Ponte* and *Purves* 1974). However, others have found that nerve section does not affect the increase in CBF during these conditions (*Bates* et al. 1976; *Hoff* et al. 1977; *Pinard* et al. 1979; *Busija* and *Heistad* 1981; *Scremin* et al. 1983). Thus, the role of the GSPN in the regulation of CBF is not settled.

Due to the uncertainty surrounding the origins of vasodilator innervation, several investigators have used pharmacological approaches to investigate the role played by cholinergic nerves in the regulation of the cerebral circulation. Topical application of atropine to cerebral arteries does not affect diameter, which suggests that resting cholinergic tone is minimal (*Kuschinsky* et al. 1974). *Scremin* and colleagues (*Rovere* et al. 1973; *Scremin* et al. 1978) have reported that atropine profoundly attenuates increases in CBF during hypercapnia. However, effectiveness of drug blockade in their studies was dependent on the level of arterial pressure. *Rovere* et al. (1973) reported that atropine attenuated cerebrovascular dilatation when arterial pressure was less than 55 mmHg. In addition, *Scremin* et al. (1978) reported that blockade was effective in limiting rises of CBF when arterial pressure was less than 70 mmHg, but had no effect at higher levels of arterial pressure. In contrast, *Kawamura* et al. (1975), *Matsuda* et al. (1978), *Busija* and *Heistad* (1982), and *Hardebo* et al. (1982) found that atropine did not alter the cerebrovascular response to hypercapnia. We demonstrated that the dose of intravenous atropine used in our experiments was effective in blocking dilatation following topical application of acetylcholine to pial arteries (*Busija* and *Heistad* 1982). Thus, it appears that cholinergic nerves do not contribute significantly to cerebrovascular dilatation during hypercapnia.

It is possible that the predominant effect of cholinergic nerves is not a direct one on vascular tone, but rather is indirect, such that cholinergic nerves modulate sympathetic effects. There is a close anatomical relationship (within 25 nm) between nonadrenergic, possibly cholinergic, terminals with adrenergic nerve endings (*Edvinsson* et al. 1972a). Because cholinergic nerves are limited to larger cerebral arteries, where sympathetic nerves predominate, this hypothesis is attractive. Activation of nicotinic receptors on sympathetic nerves decreases the release of norepinephrine from cerebral arteries (*Edvinsson* et al. 1977) and reduces the effects of sympathetic nerve stimulation on caudate blood flow in rabbits

(*Aubineau* et al. 1980). Recently, *Duckles* (1981c) has confirmed that activation of presynaptic receptors reduces stimulation-induced norepinephrine release, but in addition found that atropine, and not hexamethonium, blocked this release. This finding is consistent with the mode of cholinergic inhibition in other vascular beds (*Shepherd* and *Vanhoutte* 1981). However, during transmural electrical stimulation, when both sympathetic and cholinergic nerves are activated, atropine does not affect norepinephrine release by cerebral vessels (*Duckles* 1982a).

Cholinergic nerves may also have other effects on cerebral vessels. For example, cholinergic nerves may modulate CSF production (*Haywood* and *Vogh* 1979; *Lindvall* et al. 1978b) and may be associated with a system that affects transport of amino acids across the vessel wall (*Duckles* 1982b).

7.3 Peptidergic Nerves

It has been demonstrated by immunohistochemical techniques that cerebral arteries receive innervation by fibers containing vasoactive intestinal polypeptide (VIP) and substance P (*Chan-Palay* 1977; *Larsson* et al. 1976; *Edvinsson* et al. 1980, 1981). In addition, direct measurements demonstrate the presence of VIP in cerebral arteries (*Bevan* et al. 1982a; *Duckles* 1982a). A portion of the nonadrenergic nerves identified in cerebral arteries may be peptidergic fibers. The origins of VIP fibers are unclear, but there is a close correlation between the presence of VIP and cholinergic innervation in cats (*Bevan* et al. 1982b). Since these two neurotransmitters are released simultaneously from the same nerves to glands of the head (*Lundberg* et al. 1979; *Bloom* and *Edwards* 1980), they probably share common origins. The chorda tympani, which together with the GSPN comprise the two branches of the nervus intermedius, supply both cholinergic and VIP fibers to the salivary glands (*Lundberg* et al. 1979; *Bloom* and *Edwards* 1980). The origins of substance P fibers are better known. *Mayberg* et al. (1981) and *Liu-Chen* et al. (1983) report that these fibers arise from the trigeminal ganglia. Substance P fibers are sensory-primary afferents as indicated by depletion of this neurotransmitter after capsaicin treatment (*Duckles* and *Buck* 1982).

In vitro, VIP and substances P relax cerebral arteries (*Wei* et al. 1980d; *Edvinsson* et al. 1981; *Toda* 1982; *Winquist* et al. 1982). Similarly, topical application in vivo of these substances dilates small and large pial arteries up to 20% (*Wei* et al. 1980d; *Edvinsson* et al. 1981). Cerebral vasodilatation by VIP in vivo is blocked by cyclo-oxygenase inhibitors, which suggests that prostaglandins may be involved in the response (*Wei* et al. 1980d). Intra-arterial or intraventricular infusions of VIP increase CBF without affecting cerebral metabolism (*Heistad* et al. 1980a; *Wilson* et al. 1981). Following osmotic disruption of the blood-brain barrier, VIP increases CBF, due in part to an increase in cerebral metabolism (*McCulloch* and *Edvinsson* 1980). It is unlikely that VIP is involved in cerebral vasodilatation during hypercapnia, because CSF concentration of VIP does not change under these conditions (*Wilson* et al. 1981).

Part of the relaxation of cerebral arteries in vitro as a result of transmural stimulation, especially in cats, may be due to VIP nerves. *Bevan* et al. (1982b) have reported that VIP antibody attenuates stimulation-induced relaxation. In *Bevan's* view, both cholinergic and VIP nerves contribute to relaxation of cerebral arteries during in vitro stimulation in cats. Additional studies are needed to confirm these findings.

7.4 Central Pathways

It has been suggested that in addition to autonomic innervation via peripheral pathways, cerebral vessels also receive innervation via central nervous system pathways. In particular, it has been suggested that noradrenergic fibers that originate in the locus coeruleus and adjacent areas innervate cerebral vessels (*Hartman* et al. 1972; *Edvinsson* et al. 1973). It is unlikely that these nerves innervate cerebral resistance vessels in a conventional fashion. Electron microscopy indicates that there is always a basement membrane, associated with glial cells, between resistance vessels and nerve endings, and diffusion distances for neurotransmitters are relatively greater than in conventional neuromuscular mechanisms (*Lindvall* et al. 1975; *Rennels* et al. 1977). However, *Rennels* et al. (1977) have presented evidence which suggests that cerebral capillaries are innervated by central noradrenergic neurons.

Raichle and co-workers have reported that electrical stimulation of the locus coeruleus decreases CBF and increases capillary permeability (*Raichle* et al. 1975; *Hartman* et al. 1980). However, care must be taken in interpreting these findings to insure that brain metabolism or arterial pressure do not change during such conditions. For example, electrical stimulation of the locus coeruleus decreases the firing rate of distant neurons, reduces cerebral metabolism, and consequently decreases CBF (*Abraham* et al. 1979; *Katayama* et al. 1981). Recently, *Dahlgren* et al. (1981b) found that bilateral denervation of the locus coeruleus did not change cerebral metabolism or CBF, or influence changes in these variables during hypercapnia or hypoxia. In addition, *McCulloch* et al. (1982e) presented evidence that capillary permeability to water in the brain does not change following locus coeruleus lesions. Thus, there is inconclusive evidence that the locus coeruleus affects CBF or cerebral capillary permeability.

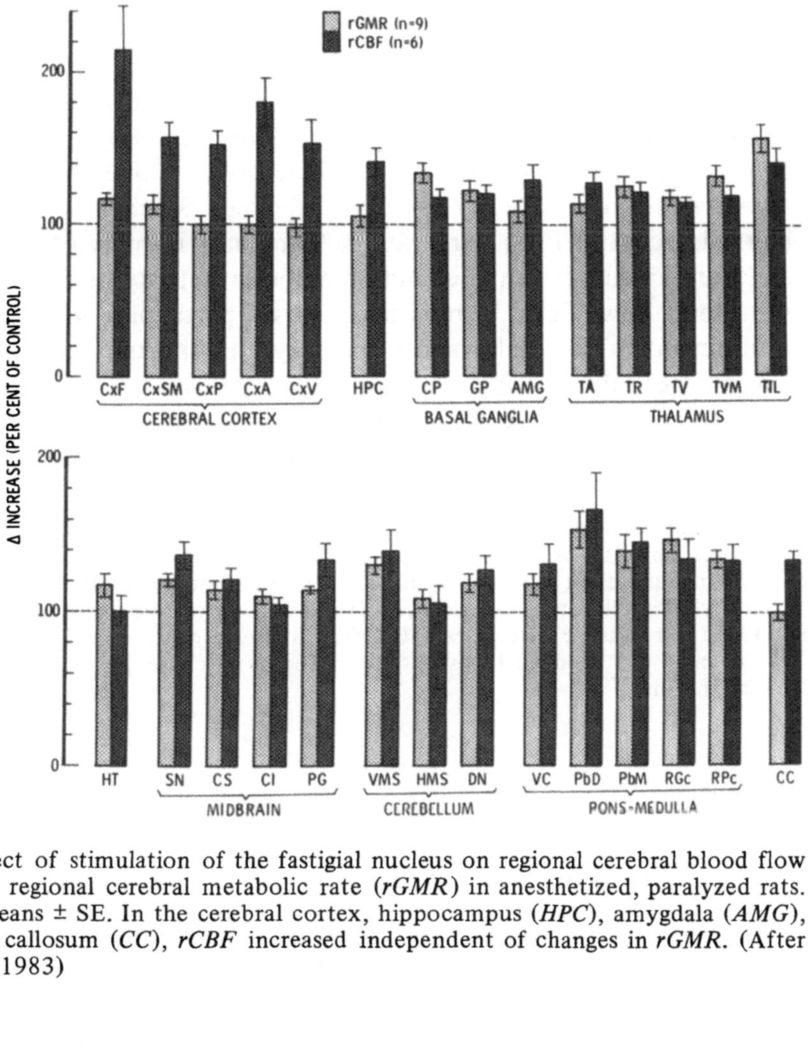

Fig. 4. Effect of stimulation of the fastigial nucleus on regional cerebral blood flow (*rCBF*) and regional cerebral metabolic rate (*rGMR*) in anesthetized, paralyzed rats. Data are means ± SE. In the cerebral cortex, hippocampus (*HPC*), amygdala (*AMG*), and corpus callosum (*CC*), *rCBF* increased independent of changes in *rGMR*. (After *Nakai* et al. 1983)

There are several provocative studies that suggest activation of other brain areas is able to affect CBF directly. *Reis* and colleagues (*Nakai* et al. 1982, 1983; *Iadecola* et al. 1982, 1983) have found that electrical activations of distinct intrinsic pathways in the brain have substantial effects on CBF. Activation of the dorsal medullary reticular formation results in widespread increases in CBF with accompanying increases in cerebral metabolism (*Iadecola* et al. 1982, 1983). In contrast, activation of the fastigial nucleus also causes widespread increases in CBF, but is unaccompanied by increases in cerebral metabolism (*Nakai* et al. 1982, 1983) (Fig. 4). Thus, central neural pathways may be important in the regulation of CBF.

8 Concluding Remarks

Although study of the regulation of the cerebral circulation has been hindered seriously, compared with other regional circulations, by methodological problems, current methods are sufficiently developed to allow accurate characterization of cerebral vascular responses. Several new concepts have emerged recently. First, large arteries are important resistance vessels in the cerebral circulation. Second, metabolic activity is matched closely with blood flow. Third, sympathetic nerves participate in the growth and development of cerebral arteries and play an important role in the regulation of CBF during hypertension, hypercapnia, hypoxia, and perhaps hypotension. Fourth, central nervous system pathways may be involved in the regulation of CBF, independent of change in cerebral metabolism.

Acknowledgments. We thank Drs. P.M. Gross and S.P. Duckles for critically reviewing the manuscript, Ms. M.L. Gray for assistance in preparing the manuscript, and Ms. M.E. Leonhauser for typing the manuscript. Original studies by the authors were supported by NRSA Individual Fellowship HL-06016, New Investigator Research Award HL-30251, Program Project Grant HL 14388, Research Grants HL 16066 and HL 30260, and by a Medical Investigatorship and research support from the Veterans Administration.

References

Abboud FM (1981) Special characteristics of the cerebral circulation. Fed Proc 40: 2296–2300

Abraham WC, Delanay RL, Dunn AJ, Zornetzer SF (1979) Locus coeruleus stimulation decreases deoxyglucose uptake in ipsilateral mouse cerebral cortex. Brain Res 172:387–392

Alm A, Bill A (1973) The effect of stimulation of the cervical sympathetic chain on retinal oxygen tension and on uveal, retinal and cerebral blood flow in cats. Acta Physiol Scand 88:84–94

Amano T, Meyer JS (1981) Prostaglandin and human CBF control: effect of aspirin and indomethacin. J Cereb Blood Flow Metab 1:S403–S404

Armstrong ML, Heistad DD, Marcus ML, Piegors DG, Abboud FM (1983) Hemodynamic sequelae of regression of experimental atherosclerosis. J Clin Invest 71:104–113

Astrup J, Heuser D, Lasser NA, Nilsson B, Nordberg K, Siesjö BK (1976) Evidence against H^+ and K^+ as the main factors in the regulation of cerebral blood flow during epileptic discharges, acute hypoxia, amphetamine intoxication, and hypoglycemia. A microelectrode study. In: Betz E (ed) Ionic actions on vascular smooth muscle. Springer, Berlin Heidelberg New York, pp 110–115

Astrup J, Heuser D, Lasser NA, Nilsson B, Nordberg K, Siesjö BK (1979) Evidence against H^+ and K^+ as main factors for the control of cerebral blood flow: A microelectrode study. Ciba Found Symp 56:313–332

Aubineau P, Sercombe R, Seylez J (1980) Parasympathetic influence of carbachol on local cerebral blood flow in the rabbit by a direct vasodilator action and an inhibition of the sympathetic mediated vasoconstriction. Br J Pharmacol 68:449–459

Auer LM, Johansson BB, Lund S (1981) Reaction of pial arteries and veins to sympathetic stimulation in the cat. Stroke 12:528–531
Barry DI, Strandgaard S, Graham DI, Brandstrup O, Svendsen UG, Vorstrup S, Hemmingsen R, Bolwig TG (1982) Cerebral blood flow in rats with renal and spontaneous hypertension: Resetting of the lower limit of autoregulation. J Cereb Blood Flow Metab 2:347–354
Barry DI, Vorstrup S, Jarden JO, Graham DI, Svendsen UG, Brandstrup O, Strandgaard S (1983) Chronic antihypertensive treatment normalizes autoregulation of cerebral blood flow in hypertension. J Cereb Blood Flow Metab 3 (Suppl 1):S656–S657
Bates DB, Sundt TM Jr (1976) The relevance of peripheral baroreceptors and chemoreceptors to regulation of cerebral blood flow in the cat. Circ Res 38:488–493
Baumbach GL, Heistad DD (1983) Effects of sympathetic stimulation and changes in arterial pressure on segmented resistance of cerebral vessels in rabbits and cats. Circ Res 52:527–533
Berne RM (1980) The role of adenosine in the regulation of coronary blood flow. Circ Res 47:807–813
Betz E (1972) Cerebral blood flow: its measurement and regulation. Physiol Rev 52: 595–630
Betz E, Csornai M (1978) Action and interaction of perivascular H^+, K^+ and Ca^{++} on pial arteries. Pflügers Arch 374:67–72
Bevan AT, Honour AJ, Stott FH (1969) Direct arterial pressure recording in unrestricted man. Clin Sci 36:329–344
Bevan JA (1979) Sites of transition between functional systemic and cerebral arteries occur at embryological functional sites. Science 204:635–637
Bevan JA (1983) Diltiazem selectively inhibits cerebrovascular extrinsic but not intrinsic myogenic tone. A review. Circ Res 52 (Suppl 1):104–108
Bevan JA, Bevan RD (1977) Sympathetic control of the rabbit basilar artery. In: Owman C, Edvinsson L (eds) Neurogenic control of the brain circulation. Pergamon, Oxford, pp 285–293
Bevan JA, Duckles SP, Lee TJF (1975) Histamine potentiation of nerve- and drug-induced responses of rabbit cerebral artery. Circ Res 36:647–653
Bevan JA, Buga GM, Florence VM, Gonsalves A, Snowden A (1982a) Distribution of choline acetyl transferase in cerebral and extracerebral cranial arteries of the cat: its relationship to neurogenic atropine-sensitive dilation. Circ Res 50:470–476
Bevan JA, Buga GM, Snowden A, Said SI (1982b) Is the neural vasodilator mechanism to cerebral and extracerebral arteries the same? In: Heistad DD, Marcus ML (eds) Cerebral blood flow: effect of nerves and neurotransmitters. Elsevier, New York, pp 421–430
Bevan RD, Tsuru H, Bevan JA (1983) Cerebral artery mass in the rabbit is reduced by chronic sympathetic denervation. Stroke 14:393–398
Bill A (1979) Effects of indomethacin on regional blood flow in conscious rabbits – a microsphere study. Acta Physiol Scand 105:437–442
Bill A, Linder J (1976) Sympathetic control of cerebral blood flow in acute arterial hypertension. Acta Physiol Scand 96:114–121
Bloom SR, Edwards AV (1980) Vasoactive intestinal peptide in relation to atropine resistant vasodilation in the submaxillary gland of the cat. J Physiol (Lond) 300: 41–53
Boisvert DPJ, Jones JV, Harper AM (1977) Cerebral blood flow autoregulation to acutely increasing blood pressure during sympathetic stimulation. In: Ingvar DH, Lassen NA (eds) Cerebral function, metabolism, and circulation. Munksgaarde, Copenhagen, pp 46–47
Borgstrom L, Johansson H, Siesjö BK (1975) The relationship between arterial PO_2 and cerebral blood flow in hypoxic hypoxia. Acta Physiol Scand 93:423–432
Bradbury MWB, Cserr HF, Westrop RJ (1981) Drainage of cerebral interstitial fluid into deep cervical lymph of rabbits. Am J Physiol 240:F329–F336

Busija DW (1983a) Role of prostaglandins in the response of the cerebral circulation to carbon dioxide in awake rabbits. J Cereb Blood Flow Metab 3:276–280
Busija DW (1983b) Effects of unilateral and bilateral sympathetic stimulation on cerebral blood flow in rabbits. Physiologist 26:A92 (abstract)
Busija DW (1984a) Sympathetic nerves reduce cerebral blood flow during hypoxia in awake rabbits. Am J Physiol 247 (Sept) (to be published)
Busija DW (1984b) Role of prostaglandins in modulating sympathetic vasoconstriction in the cerebral circulation in anesthetized rabbits. J Cereb Blood Flow Metab (to be published)
Busija DW, Heistad DD (1981) Effects of cholinergic nerves on cerebral blood flow in cats. Circ Res 48:62–69
Busija DW, Heistad DD (1982) Atropine does not attenuate cerebral vasodilation during hypercapnia. Am J Physiol 11:H683–H687
Busija DW, Heistad DD (1983) Effects of indomethacin on cerebral blood flow during hypercapnia in cats. Am J Physiol 244:H519–H524
Busija DW, Heistad DD (1984) Effects of activation of sympathetic nerves on cerebral blood flow during hypercapnia in cats and rabbits. J Physiol (Lond) 347:35–45
Busija DW, Heistad DD, Marcus ML (1980a) Effects of sympathetic nerves on cerebral vessels during acute, moderate increases in arterial pressure in dogs and cats. Circ Res 46:696–702
Busija DW, Orr JA, Rankin JHG, Liang HK, Wagerle LC (1980b) Cerebral blood flow during normocapnic hyperoxia in the unanesthetized pony. J Appl Physiol 48:10–15
Busija DW, Heistad DD, Marcus ML (1981) Continuous measurement of cerebral blood flow in anesthetized cats and dogs. Am J Physiol 241:H228–H234
Busija DW, Marcus ML, Heistad DD (1982a) Effects of sympathetic nerves on pial artery diameter and blood flow velocity in cats. J Cereb Blood Flow Metab 2:363–367
Busija DW, Sadoshima S, Marcus ML, Heistad DD (1982b) Functional significance of sympathetic innervation of cerebral vessels. Can the issue now be resolved? In: Kalsner S (ed) Trends in autonomic pharmacology. Urban and Schwarzenberg, Baltimore, pp 187–204
Cameron IR, Caronna J (1976) The effect of local changes in potassium and bicarbonate concentration on hypothalamic blood flow in the rabbit. J Physiol (Lond) 262:415–430
Carpenter MB (1978) Core text of neuroanatomy. Williams and Wilkins, Baltimore, pp 317–340
Caspers H, Speckmann EJ (1972) Cerebral pO_2, pCO_2 and pH: Changes during convulsive activity and their significance for spontaneous arrest of seizures. Epilepsia 13: 699–725
Chan-Palay V (1977) Innervation of cerebral blood vessels by norepinephrine, indoleamine, substance P and neurotensin fibers and the leptomeningeal indoleamine axons: their roles in vasomotor activity and local alterations of brain blood composition. In: Owman C, Edvinsson L (eds) Neurogenic control of brain circulation. Pergamon, Oxford, pp 39–53
Chapman AG, Meldruni BS, Siesjö BK (1977) Cerebral metabolic changes during prolonged epileptic seizures in rats. J Neurochem 28:1025–1035
Chorobski J, Penfield W (1932) Cerebral vasodilator nerves and their pathway from the medulla oblongata. Arch Neurol Psychiatry 28:1257–1289
Cobb S, Finesinger JE (1932) Cerebral circulation XIX. The vagal pathway of the vasodilator impulse. Arch Neurol Psychiatry 28:1234–1256
Craigen ML, Jennett S (1981) Pial arterial response to systemic hypoxia in anesthetized cats. J Cereb Blood Flow Metab 1:285–296
Crockard HA, Iannotti F, Ladds G (1982) Cerebrovascular effects of prostaglandin inhibitors in the gerbil. J Cereb Blood Flow Metabol 2:67–72
Dahlgren N, Ingvar M, Yokayama H, Siesjö BK (1981a) Effect of indomethacin on local cerebral blood flow in awake, minimally restrained cats. J Cereb Blood Flow Metab 1:233–236

Dahlgren N, Lindvall O, Sakabe T, Steneni U, Siesjö BK (1981b) Cerebral blood flow and oxygen consumption in the rat brain after lesions of the noradrenergic locus coeruleus system. Brain Res 209:11–23
Dahlgren N, Nilsson B, Sakabe T, Siesjö BK (1981c) The effect of indomethacin on cerebral blood flow and oxygen consumption in the rat at normal and increased carbon dioxide tensions. Acta Physiol Scand 111:475–485
D'Alecy LG, Feigl EO (1972) Sympathetic control of cerebral blood flow in dogs. Circ Res 31:267–283
D'Alecy LG, Rose CJ (1977) Parasympathetic cholinergic control of cerebral blood flow in dogs. Circ Res 41:324–331
D'Alecy LG, Rose CJ, Sellers SA (1979) Sympathetic modulation of hypercapnic cerebral vasodilation in dogs. Circ Res 45:771–785
Davson H (1967) Physiology of the cerebrospinal fluid. Churchill, London, pp 183–184
De La Torre JC, Mullen S (1971) Experimental evaluation of L-dopa penetration in brain. Trans Am Neurol Assoc 96:227–229
Des Rosiers MH, Kennedy C, Sakurada O, Shinohara M, Sokoloff L (1978) Effects of hypercapnia on cerebral oxygen and glucose consumption in the conscious rat. Stroke 9:98 (Abstract)
Diepstra G, Gonylea W, Mitchell JH (1980) Cardiovascular response to static exercise during selective autonomic blockade in the conscious cat. Circ Res 47:530–535
Duckles SP (1979) Neurogenic dilator and constrictor responses of pial arteries in vitro. Differences between dogs and sheep. Circ Res 44:482–490
Duckles SP (1981a) Vasodilator innervation of cerebral blood vessels. In: Vanhoutte PM, Leusen I (eds) Vasodilatation. Raven, New York, pp 27–38
Duckles SP (1981b) Evidence for a functional cholinergic innervation of cerebral arteries. J Pharmacol Exp Ther 217:544–548
Duckles SP (1981c) Evidence against a physiological role for presynaptic acetylcholine receptors on cerebrovascular adrenergic nerves. Fed Proc 40:421 (Abstract)
Duckles SP (1982a) Acetylcholine and vasoactive intestinal polypeptide: Cerebrovascular neurotransmitters? In: Heistad DD, Marcus ML (eds) Cerebral blood flow: effects of nerves and neurotransmitters. Elsevier North Holland, New York, pp 441–446
Duckles SP (1982b) Choline acetyltransferase in cerebral arteries: modulator of amino acid uptake? J Pharmacol Exp Ther 223:716–720
Duckles SP, Bevan JA (1976) Pharmacological characterization of adrenergic receptors of a rabbit cerebral artery in vitro. J Pharmacol Exp Ther 197:371–378
Duckles SP, Bevan JA (1979) Responses of small rabbit pial arteries in vitro. Blood Vessels 16:80–86
Duckles SP, Buck SH (1982) Substance P in the cerebral vasculature: depletion by capsaicin suggests a sensory role. Brain Res 245:171–174
Duckles SP, Rapoport R (1979) Release of endogenous norepinephrine from a rabbit cerebral artery. J Pharmacol Exp Ther 211:219–224
Duckles SP, Lee TJF, Bevan JA (1977) Cerebral arterial responses to nerve stimulation in vitro. Species variation in the constrictor and dilator components. In: Owman C, Edvinsson L (eds) Neurogenic control of the brain circulation. Pergamon, Oxford, pp 133–142
Duffy TE, Carazzuti M, Cruz NF, Sokoloff L (1982) Local cerebral glucose metabolism in newborn dogs: effects of hypoxia and halothane anesthesia. Ann Neurol 11:233–246
Edvinsson L (1975) Neurogenic mechanism in the cerebrovascular bed. Autonomic nerves, amine receptors and their effects in cerebral blood flow. Acta Physiol Scand (Suppl) 427:1–36
Edvinsson L, MacKenzie ET (1977) Amine mechanism in the cerebral circulation. Pharmacol Rev 28:275–348

Edvinsson L, Owman Ch (1974) Pharmacological characterization of adrenergic alpha and beta receptors mediating the vasomotor response of cerebral arteries in vitro. Circ Res 35:835–849

Edvinsson L, Owman Ch (1977) Sympathetic innervations and adrenergic receptors in intraparenchymal cerebral arteries of baboon. In: Ingvar DH, Lasser NA (eds) Proceedings of the 8th international symposium on cerebral function, metabolism and circulation. Munksgaarde, Copenhagen, pp 304–305

Edvinsson L, Owman C, West KA (1971) Changes in cerebral blood volume of mice at various time-periods after superior cervical sympathectomy. Acta Physiol Scand 82: 521–526

Edvinsson L, Nielsen KC, Owman CH, Sporrong B (1972a) Cholinergic mechanisms in pial vessels. Z Zellforsch 134:311–325

Edvinsson L, Owman Ch, Rosengrent E, West KA (1972b) Concentration of noradrenaline in pial vessels, choroid plexus, and iris during 2 weeks after sympathetic ganglionectomy or decentralization. Acta Physiol Scand 85:201–206

Edvinsson L, Lindvall M, Nielsen KC, Owman C (1973) Are brain vessels innervated also by central (non-sympathetic) adrenergic neurones? Brain Res 63:396–499

Edvinsson L, Nielsen KC, Owman C, West KA (1974) Adrenergic innervation of the mammalian choroid plexus. Am J Anat 139:299–308

Edvinsson L, Hakanson M, Lindvall M, Owman Ch, Svensson KG (1975) Ultrastructural and biochemical evidence for a sympathetic neural influence on the choroid plexus. Exp Neurol 48:241–251

Edvinsson L, Owman C, Siesjö B (1976a) Physiological role of cerebrovascular sympathetic nerves in the autoregulation of cerebral blood flow. Brain Res 117:518–523

Edvinsson L, Owman C, Sjoberg NO (1976b) Autonomic nerves, mast cells and amine receptors in human brain vessels; A histochemical and pharmacological study. Brain Res 115:337–393

Edvinsson L, Falck B, Owman Ch (1977) Possibilities for a cholinergic action on smooth musculature and on sympathetic axons in brain vessels mediated by muscarinic and nicotinic receptors. J Pharmacol Exp Ther 200:117–126

Edvinsson L, Fahrenkrug J, Hanko J, Owman Ch, Sundler F, Uddman R (1980) VIP nerve supply of intracranial arteries of mammals. Cell Tissue Res 208:135–142

Edvinsson L, McCulloch J, Uddman R (1981) Substance P: immunohistochemical localization and effect upon pial arteries in vitro and in situ. J Physiol (Lond) 318: 251–258

Eklöf B, MacMillan V, Siesjö BK (1972) Cerebral energy state and cerebral venous pO_2 in experimental hypotension caused by bleeding. Acta Physiol Scand 86:515–527

Ellis EF, Wei EP, Cockrell CS, Traweek DL, Saady JJ, Kontos HA (1982) The effect of O_2 and CO_2 on prostaglandin levels in the cat cerebral cortex. Circ Res 51:652–656

Emerson TB Jr, Raymond RM (1981) Involvement of adenosine in cerebral hypoxic hyperemia in the dog. Am J Physiol 241:H134–H138

Eriksson S, Hagenfeldt L, Law D, Patrono C, Pinca E, Wennmalm A (1983) Effect of prostaglandin synthesis inhibitors on basal and CO_2-stimulated cerebral blood flow in man. Acta Physiol Scand 117:203–211

Farrar JK, Jones JV, Graham DI, Strandgaard S, MacKenzie ET (1976) Evidence against cerebral vasospasm during acutely induced hypertension. Brain Res 104: 176–180

Fitch W, MacKenzie ET, Hasper AM (1975) Effects of decreasing arterial blood pressure in cerebral blood flow in the baboon. Circ Res 37:550–557

Florence VM, Bevan JA (1979) Biochemical determinations of cholinergic innervation in cerebral arteries. Circ Res 45:217–218

Flower RJ (1974) Drugs which inhibit prostaglandin biosynthesis. Pharmacol Rev 26: 33–67

Folkow B, Neil E (1971) Circulation. Oxford University Press, New York, p 46

Foreman DL, Sanders M, Bloor CM (1976) Total and regional cerebral blood flow during moderate and severe exercise in miniature swine. J Appl Physiol 40:191–195

Fujiwara S, Itoh T, Suzuki H (1982) Membrane properties and excitatory neuromuscular transmission in the smooth muscle of dog cerebral arteries. Br J Pharmacol 77: 197–208

Futamachi KJ, Mutani R, Prince DA (1974) Potassium activity in rabbit cortex. Brain Res 75:5–25

Galvan M, Ten Bruggencate G, Senekowitsch R (1979) The effects of neuronal stimulation and ouabain upon extracellular K^+ and C^{2+} levels in rat isolated sympathetic ganglia. Brain Res 160:544–548

Gotoh F, Tazaki Y, Meyer JS (1961) Transport of gases through brain and their extravascular vasomotor action. Exp Neurol 4:48–58

Gotoh F, Muramatsu F, Fukuuchi Y, Okayasu H, Tanaka K, Suzuki N, Kobari M (1982) Video camera method for simultaneous measurement of blood flow velocity and pial vessel diameters. J Cereb Blood Flow Metab 2:421–428

Graham DI, Grome JJ, Kelly PAT, MacKenzie ET, McCulloch J, Reis DJ, Talman WT (1982) Cerebral circulatory effect of fulminating neurogenic hypertension. In: Heistad DD, Marcus ML (eds) Cerebral blood flow: effects of nerves on neurotransmitters. Elsevier North Holland, New York, pp 493–502

Gross PM, Heistad DD, Strait MR, Marcus ML, Brody MJ (1979) Cerebral vascular responses to physiological stimulation of sympathetic pathways in cats. Circ Res 44:288–294

Gross PM, Marcus ML, Heistad DD (1980) Regional distribution of cerebral blood flow during exercise in dogs. J Appl Physiol 48:213–217

Gross PM, Harper AM, Teasdale GM (1981a) Cerebral circulation and histamine 1. Participation of vascular H_1- and H_2-receptors in vasodilatatory responses to carotid arterial infusion. J Cereb Blood Flow Metabol 1:97–108

Gross PM, Harper AM, Grahm DI (1981b) Cerebral blood flow in rats during physiological and humoral stimuli. Stroke 12:345–352

Gross PM, Reasdale GM, Angersen WJ, Harper AM (1982) Intra-arterial histamine increases blood-brain transport in rats. Am J Physiol 234:H307–H317

Gross PM, Harper AM, Reasdale GM (1983) Interaction of histamine with noradrenergic constrictory mechanisms in cat cerebral arteries and veins. Can J Physiol Pharmacol 61:756–763

Häggendal E, Johansson B (1965) Effects of arterial carbon dioxide tension and oxygen saturation on cerebral blood flow autoregulation in dogs. Acta Physiol Scand 66 (Suppl 2–8):27–53

Hanley DF, Wilson DA, Traystman RJ (1983) Cerebral regions without hypoxic or hypercapnic responses. J Cereb Blood Flow Metab 3:S632–S633

Hardebo JE (1981) Vasodilation augments the blood-brain barrier lesions induced by an acute rise in introcarotid pressure. Blood Vessels 18:9–15

Hardebo JE, Owman C (1980) Barrier mechanisms for neurotransmitter monoamines and their precursors at the blood-brain interface. Ann Neurol 8:1–11

Hardebo JE, Lindvall O, Nilsson B (1982) On the possible influence of adrenergic and cholinergic mechanisms in normo- and hypercapnia. In: Heistad DD, Marcus ML (eds) Cerebral blood flow: effect of nerves and neurotransmitters. Elsevier, New York, pp 377:384

Harder DR (1980) Comparison of electrical properties of middle cerebral and mesenteric artery in the cat. Am J Physiol 239:C23–C26

Harder DR, Abel PW, Hermsmeyer K (1981) Membrane electrical mechanism of basilar artery constriction and pial artery dilation by norepinephrine. Circ Res 49:1237–1242

Harper AM (1966) Autoregulation of cerebral blood flow: influence of the arterial blood pressure on the blood flow through the cerebral cortex. J Neurol Neurosurg Psychiatry 29:398–403

Harper AM, Bell RA (1963) The effect of metabolic acidosis and alkalosis on the blood flow through cerebral cortex. J Neurol Neurosurg Psychiatry 26:341–344
Harper AM, Deshmukh WD, Rowman JO, Jennett WB (1972) The influence of sympathetic nervous activity on cerebral blood flow. Arch Neurol 27:1–6
Hart M, Heistad DD, Brody MJ (1980) Effect of chronic hypertension and sympathetic denervation on wall-lumen ratio of cerebral arteries. Hypertension 2:419–423
Hartman BK, Zide D, Udenfriend D (1972) The use of dopamine beta hydroxylase as a marker for the central noradrenergic nervous system in rat brain. Proc Natl Acad Sci USA 69:2722–2726
Hartman BK, Swanson LW, Raichle ME, Preskorn SH, Clark HB (198) Central adrenergic regulation of cerebral microvascular permeability and blood flow; anatomic and physiologic evidence. Adv Exp Biol Med 131:113–126
Haywood JR, Vogh BP (1979) Some measurements of autonomic nervous system influence on production of cerebrospinal fluid in the cat. J Pharmacol Exp Ther 208–341–346
Hegedus SA, Shackelford RT (1965) A comparative-anatomical study of the cranial-cervical venous systems in mammals, with special reference to the dog: relationship of anatomy to measurements of cerebral blood flow. Am J Anat 116:375–386
Heinemann U, Lux HD, Gutnick MJ (1977) Extracellular free calcium and potassium during paroxysmal activity in the cerebral cortex of the cat. Exp Brain Res 27: 237–243
Heistad DD, Marcus ML (1976) Total and regional cerebral blood flow during stimulation of carotid baroreceptors. Stroke 7:239–243
Heistad DD, Marcus ML (1978) Evidence that neural mechanisms do not have important effects on cerebral blood flow. Circ Res 42:295–302
Heistad DD, Marcus ML (1979) Effect of sympathetic stimulation on permeability of the blood-brain barrier to albumin during acute hypertension in cats. Circ Res 45: 331–338
Heistad DD, Marcus ML, Ehrhardt JC, Abboud FM (1976) Effect of stimulation of carotid chemoreceptors on total and regional recebral blood flow. Circ Res 38:20–25
Heistad DD, Marcus ML, Sandberg S, Abboud FM (1977) Effect of sympathetic nerve stimulation on cerebral blood flow and on large cerebral arteries of dog. Circ Res 41:342–350
Heistad DD, Marcus ML, Abboud FM (1978a) Role of large arteries in regulation of cerebral blood flow in dogs. J Clin Invest 62:761–768
Heistad DD, Marcus ML, Gross PM (1978b) Effects of sympathetic nerves on cerebral vessels in dog, cat, and monkey. Am J Physiol 235:H544–H552
Heistad DD, Marcus ML, Said SI, Gross PM (1980a) Effect of acetylcholine and vasoactive intestinal peptide on cerebral blood flow. Am J Physiol 239:H73–H80
Heistad DD, Marcus ML, Piegors DJ, Armstrong ML (1980b) Regulation of cerebral blood flow in atherosclerotic monkeys. Am J Physiol 239:H539–H544
Heistad DD, Gross PM, Busija DW, Marcus ML (1980c) Cerebral vascular response to loading and unloading of arterial baroreceptors. In: Sleight P (ed) Arterial baroreceptors and hypertension. Oxford, Oxford, pp 210–217
Heistad DD, Marcus ML, Gourley JK, Busija DW (1981) Effect of adenosine and dipyridamole on cerebral blood flow. Am J Physiol 9:H775–H780
Hernandez MJ, Brennan RW, Bowman GS (1980) Autoregulation of cerebral blood flow in the newborn dog. Brain Res 184:199–202
Heuser D (1979) The significance of cortical extracellular H^+, K^+ and Ca^{2+} activities for regulation of local cerebral blood flow under conditions of enhanced neuronal activity. In: Purves MJ (ed) Cerebral vascular smooth muscle and its control. Elsevier, Amsterdam, pp 339–349
Hoff JT, MacKenzie ET, Harper AM (1977) Responses of the cerebral circulation of hypercapnia and hypoxia after 7th cranial nerve transection in baboons. Circ Res 40:258–262

Iadecola C, Arbit E, Nakai M, Mraovitch S, Tucker L, Reis DJ (1982) Increased regional cerebral blood flow and metabolism electricity by stimulation of the dorsal medullary reticular formation in the rat: evidence for an intrinsic neural system in brain regulating cerebral metabolism. In: Heistad DD, Marcus ML (eds) Cerebral blood flow: effects of nerves and neurotransmitters. Elsevier, North Holland, pp 485–492

Iadecola C, Nakai M, Arbit E, Reis DJ (1983) Global cerebral vasodilatation elicited by focal electrical stimulation within the dorsal medullary reticular formation in anesthetized rat. J Cereb Blood Flow Metab 3:270–279

Ingvar DH, Lassen NA (1982) Regional blood flow of the cerebral cortex determined by Krypton^{85}. Acta Physiol Scand 54:325–338

Jackson EK, Gerkens JF, Zimmerman JB, Uderman HD, Oates JA, Workman RJ, Branch RA (1983) Prostaglandin biosynthesis does not participate in hypercapnia-induced cerebral vasodilation in the dog. J Pharmacol Exp Ther 226:486–492

James IM, MacDonell L (1975) The role of baroreceptors and chemoreceptors in the regulation of the cerebral circulation. Clin Sci 49:465–471

James IM, Millar RA, Purves MJ (1969) Observations on the extrinsic neural control of cerebral blood flow in the baboon. Circ Res 25:77–93

Johansson B, Mellander S (1975) Static and dynamic componenty in the vascular response to passive changes in length as revealed by electrical and mechanical recordings from the rat portal vein. Circ Res 36:76–83

Johansson BB, Linder LE (1980) The blood brain barrier in renal hypertensive rats. Clin Exp Hypertension 2(6):983–993

Johansson BB, Nordborg C (1978) Cerebral vessels in spontaneously hypertensive rat. Adv Neurol 20:349–357

Johnson PC, Henrich HA (1975) Metabolic and myogenic factors in local regulation of the microcirculation. Fed Proc 34:2020–2024

Jones JV, Fitch W, MacKenzie ET, Strandgaard S, Harper AM (1976) Lower limit of cerebral blood flow and autoregulation in experimental renovascular hypertension in the baboon. Circ Res 39:555–557

Jones MD Jr, Sheldon RE, Peeters LL, Makoski EL, Meschia G (1978) Regulation of cerebral blood flow in the ovine fetus. Am J Physiol 235:H162–H166

Jones MD Jr, Traystman RJ, Simmons MA, Molteni RA (1981) Effects of changes in arterial O_2 content on cerebral blood flow in the lamb. Am J Physiol 240:H209–H215

Jonsson O (1970) Extracellular osmolality and vascular smooth muscle activity. Acta Physiol Scand (Suppl) 359:1–48

Katayama Y, Aeno Y, Tsukiyama T, Tsubakawa T (1981) Long lasting suppression of firing of cortical neurons and decrease in cortical blood flow following train pulse stimulation of the locus coeruleus in the cat. Brain Res 216:173–179

Kato M, Ueno H, Black P (1974) Regional cerebral blood flow of the main visual pathways during photic stimulation of the retina in intact and split-brain monkeys. Exp Neurol 42:65–77

Kawamura Y, Meyer JS, Hiromoto H, Aoyagi M, Tagashira Y, Ott EO (1975) Neurogenic control of cerebral blood flow in the baboon. J Neurosurg 43:676–688

Koehler RC, Traystman RJ (1982) Bicarbonate ion modulation of cerebral blood flow during hypoxia and hypercapnia. Am J Physiol 243:H33–H40

Kogure K, Scheinberg P, Reinmuth OM, Fujishima M (1970a) Effects of hypoxia on cerebral autoregulation. Am J Physiol 219:1393–1396

Kogure K, Scheinberg P, Reinmuth OM, Fujishima M, Busto R (1970b) Mechanisms of cerebral vasodilatation in hypoxia. J Appl Physiol 29:223–229

Kontos HA, Raper AJ, Patterson JL Jr (1977a) Analysis of vasoactivity of local pH, pCO_2 and bicarbonate and pial vessels. Stroke 8:358–360

Kontos HA, Wei EP, Raper AJ, Patterson JL Jr (1977b) Local mechanism of CO_2 action on cat pial arterioles. Stroke 8:226–229

Kontos HA, Wei EP, Navari RM, Levasseur JE, Rosenblum WI, Patterson JL Jr (1978a) Responses of cerebral arteries and arterioles to acute hypotension and hypertension. Am J Physiol 234:H371–H383

Kontos HA, Wei EP, Raper AJ, Rosenblum WI, Navari RM, Patterson JL Jr (1978b) Role of tissue hypoxia in local regulation of cerebral microcirculation. Am J Physiol 234:H582–H591

Kreutzberg GW, Barron KD, Schubert P (1978) Cytochemical localization of 5′-nucleotidose in glial plasma membranes. Brain Res 158:247–257

Krishner HS, Blank WF, Myers RE (1975) Brain extracellular potassium activity during hypoxia in the cat. Neurology 25:1001–1005

Krishner HS, Blank WF Jr, Myers RE (1976) Changes in cortical subarachnoid fluid potassium concentrations during hypoxia. Arch Neurol 33:84–90

Kuschinsky W, Wahl M (1975) Alpha-receptor stimulation by endogenous and exogenous norepinephrine and blockade by phentolamine in pial arteries in cats. Circ Res 37:168–174

Kuschinsky W, Wahl M (1979) Perivascular pH and pial arterial diameter during bicuculline induced seizures in cats. Pflügers Arch 382:81–85

Kuschinsky W, Wahl M, Bosse O, Thurau K (1972) Perivascular potassium and pH as determinants of local pial arterial diameter in cats. A microapplication study. Circ Res 31:240–247

Kuschinsky W, Wahl M, Weiss A (1974) Evidence for cholinergic dilatory receptors in pial arteries of cat: a microapplication study. Pflügers Arch 347:199–208

Kuschinsky W, Suda S, Sokoloff L (1981) Local cerebral glucose utilization and blood flow during metabolic acidosis. Am J Physiol 241:H772–H777

Lacombe P, Reynier-Rebuffel AM, Masmo H, Seylaz J (1977) Quantitative multiregional blood flow measurements during cervical sympathetic stimulation. Brain Res 129:129–140

Langer AW, Obrist PA, McCubbin JA (1979) Hemodynamic and metabolic adjustments during exercise and shock avoidance in dogs. Am J Physiol 236:H225–H230

Langfitt TW, Weinstein JD, Kassell NF (1966) Vascular factors in head injury. In: Caveness WF, Walker AE (eds) Head injury. Lippincott, Philadelphia, pp 172–194

Lambertsen CJ, Kough RH, Cooper DY, Emmel GL, Loeschcke HH, Schmidt CF (1953) Oxygen toxicity. Effects in man of oxygen inhalation at 1 and 3.5 atmospheres upon blood gas transport, cerebral circulation and cerebral metabolism. J Appl Physiol 5:471–486

Larsson LI, Edvinsson L, Fahrenkrug J, Hakanson R, Owman C, Schaffalitzky De Musckadell O, Sundler F (1976) Immunohistochemical localization of a vasodilatory polypeptide (VIP) in cerebrovascular nerves. Brain Res 113:400–404

Lassen NA (1968) Brain extracellular pH: the main factor controlling cerebral blood flow. Scand J Clin Lab Invest 22:247–251

Lassen NA, Ingvar DH, Skinhøj E (1978) Brain function and blood flow. Sci Am 239: 62–71

Lee TJF (1980) Direct evidence against acetylcholine as the dilator transmitter in the cat cerebral artery. Eur J Pharmacol 68:393–394

Lee TJF (1982) Cholinergic mechanism in the large cat cerebral artery. Circ Res 50: 870–879

Lee TJF, Su C, Bevan JA (1976) Neurogenic sympathetic vasoconstriction of the rabbit basilar artery. Circ Res 39:120–126

Lee RJF, Hume WR, Su C, Bevan JA (1978) Neurogenic vasodilation of cat cerebral arteries. Circ Res 42:535–542

Lee TJF, Chiang CC, Adams M (1980) Synaptic transmission of vasoconstrictor nerves in rabbit basilar artery. Eur J Pharmacol 61:55–70

Leninger-Follert E, Hossman K-A (1979) Simultaneous measurements of microflow and evoked potentials in the somatomotor cortex of the cat brain during specific sensory activation. Pflügers Arch 380:85–89

Leninger-Follert E, Lübbers DW (1976) Behavior of microflow and local PO_2 of the brain cortex during and after direct electrical stimulation. Pflügers Arch 366:39–44
Levasseur JE, Wei EP, Raper AJ, Kontos HA, Patterson JL Jr (1975) Detailed description of a cranial window technique for acute and chronic experiments. Stroke 6: 308–317
Linder J (1981) Effects of facial nerve section and stimulation on cerebral and ocular blood flow in hemorrhagic hypotension. Acta Physiol Scand 112:185–193
Lindvall M, Owman C (1981) Autonomic nerves in the mammalian choroid plexus and their influence on the formation of cerebrospinal fluid. J Cereb Blood Flow Metab 1:245–266
Lindvall M, Cervos-Navarro J, Edvinsson L, Owman CH, Stenevi U (1975) Nonsympathetic perivascular nerves in the brain: Origin and mode of innervation studied by fluorescence and electronic microscopy combined with stereotaxic lesions and sympathectomy. In: Harper AM, Jennett WB, Miller JD, Rowan JO (eds) Blood flow and metabolism in the brain. Churchill-Livingstone, Edinburgh, pp 1.7–1.9
Lindvall M, Edvinsson L, Owman C (1978a) Sympathetic nervous control of cerebrospinal fluid production from the choroid plexus. Science 201:176–178
Lindvall M, Edvinsson L, Owman C (1978b) Reduced cerebrospinal fluid formation through cholinergic mechanisms. Neurosci Lett 10:311–316
Liu-Chen LY, Han DH, Moskowitz MA (1983) Pia arachnoid contains substance P originating from trigeminal neurons. Neuroscience 9:803–808
Lundberg JM, Hökfelt T, Schultzberg M, Üvnas-Wallenstein K, Köhler C, Said SI (1979) Occurrence of vasoactive intestinal polypeptide (VIP)-like immunoreactivity in certain cholinergic neurons of the cat: evidence from combined immunohistochemistry and acetylcholinesterase staining. Neuroscience 4:1539–1559
Lusamvuku NAT, Sercombe R, Aubineau P, Seylaz J (1979) Correlated electrical and mechanical responses of isolated rabbit pial arteries to some vasoactive drugs. Stroke 10:727–732
MacKenzie ET, McCulloch J, O'Keane M, Pickard J-D, Harper AM (1976a) Cerebral circulation and norepinephrine: relevance of the blood-brain barrier. Am J Physiol 231:483–488
MacKenzie ET, Strandgaard S, Graham DI, Jones JV, Harper AM, Farrar JK (1976b) Effects of acutely induced hypertension in cats on pial arteriolar caliber, local cerebral blood flow, and the blood-brain barrier. Circ Res 39:33–41
MacKenzie ET, McGeorge AP, Graham DI, Fitch W, Edvinsson L, Harper AM (1979) Effects of increasing arterial pressure on cerebral blood flow in the baboon: influence of the sympathetic nervous system. Pflügers Arch 378:189–195
Marcus ML, Heistad DD (1979) Effects of sympathetic nerves on cerebral blood flow in awake dogs. Am J Physiol 5:H549–H553
Marcus ML, Heistad DD, Ehrhardt JC, Abboud FM (1976) Total and regional cerebral blood flow measurement with 7-10-, 15-, 25-, and 50-μm microspheres. J Appl Physiol 40:501–507
Marcus ML, Bischof CJ, Heistad DD (1981a) Comparison of microsphere and Xenon-133 clearance method in measuring skeletal muscle and cerebral blood flow. Circ Res 48:748–761
Marcus ML, Busija DW, Bischof CJ, Heistad DD (1981b) Methods for measurement of cerebral blood flow. Fed Proc 40:2306–2310
Marcus ML, Busija DW, Gross PM, Brooks LA, Heistad DD (1982) Sympathetic escape in the cerebral circulation during normotension and acute severe hypertension. In: Heistad DD, Marcus ML (eds) Cerebral blood flow: effect of nerves and neurotransmitters. Elsevier, New York, pp 281–289
Mata M, Fink DJ, Gainer H, Smith CB, Davidsen L, Saraki H, Schwartz WJ, Sokoloff L (1980) Activity-dependent energy metabolism in rat posterior pituitary primarily reflects sodium pump activity. J Neurochem 34:213–215
Matsuda M, Yoneda S, Gotoh H, Handa J, Handa H (1978) Effect of atropine on cerebrovascular responsiveness to carbon dioxide. J Neurosurg 48:417–422

Mayberg M, Langer RS, Zervas NT, Moskowitz MA (1981) Perivascular meningeal projections from cat trigeminal ganglia: Possible pathway for vascular headaches in man. Science 213:228–230
McCalden TA, Bevan JA (1981) Sources of activator calcium in rabbit basilar artery. Am J Physiol 10:H129–H133
McCalden TA, Nath R, Thiele K (1982) The role of prostaglandins in the cerebrovascular dilation with hypercapnia and hypoxia. Fed Proc 41:1610 (Abstract)
McCulloch J, Edvinsson L (1980) The effects of vasoactive intestinal polypeptide upon pial arteriolar calibre, cerebral blood flow, cerebral oxygen consumption and the electroencephalogram. Am J Physiol 238:H449–H456
McCulloch J, Savaki HE, McCulloch MC, Jehle J, Sokoloff L (1982a) The distribution of alterations in energy metabolism in the rat brain produced by apomorphine. Brain Res 243:67–80
McCulloch J, Savaki HE, Sokoloff L (1982b) Distribution of effects of haloperidol on energy metabolism in the rat brain. Brain Res 243:81–90
McCulloch J, Kelly PAT, Ford I (1982c) Effect of apomorphine on the relationship between local cerebral glucose utilization and local cerebral blood flow. J Cereb Blood Flow Metab 2:487–499
McCulloch J, Kelly PAT, Grome JJ, Pickard JD (1982d) Local cerebral circulatory and metabolic effects of indomethacin. Am J Physiol 243:H416–H423
McCulloch J, Savaki HE, Angersen W (1982e) Regional water permeability in the CNS of conscious rats: Effects of hypercapnia and locus coeruleus lesions. In: Heistad DD, Marcus ML (eds) Cerebral blood flow: effects of nerves and neurotransmitters. Elsevier North Holland, New York, pp 509–516
Meldrum BS, Nilsson B (1976) Cerebral blood flow and metabolic rate early and late in prolonged epileptic seizures induced in rats by bucuculline. Brain 99:523–542
Metzger H (1979) Effects of direct stimulation on cerebral cortex oxygen tension level. Microvasc Res 17:80–89
Miller AL, Hawkins RA, Veech RL (1975) Decreased rate of glucose utilization by rat brain in vivo after exposure to atmospheres containing high concentrations of CO_2. J Neurochem 25:553–558
Miyaoka M, Shinohara M, Kennedy C, Sokoloff L (1980) Alterations in local cerebral glucose utilization (LCGU) in rat brain during hypoxemia. Trans Am Neurol Assoc 104:1–4
Morii S, Winn HR, Berne RM (1983) Effect of theophylline, an adenosine receptor blocker, on cerebral blood flow (CBF) during rest and transient hypoxia. J Cereb Blood Flow Metab 3 (Suppl 1):S480–S481
Morris ME (1974) Hypoxia and extracellular potassium activity in the guinea-pig cortex. Can J Physiol Pharmacol 52:872–882
Mueller SM, Heistad DD (1980) Effect of chronic hypertension on the blood-brain barrier. Hypertension 2:809–812
Mueller SM, Luft FC (1982) The blood-brain barrier in renovascular hypertension. Stroke 13(2):229–234
Mueller SM, Heistad DD, Marcus ML (1977) Total and regional cerebral blood flow during hypotension, hypertension and hypocapnia: Effect of sympathetic denervation in dogs. Circ Res 41:350–356
Mueller SM, Heistad DD, Marcus ML (1979) Effect of sympathetic nerves on cerebral vessels during seizures. Am J Physiol 237:H178–H184
Mueller SM, Ertel PJ, Felten DL, Overhage JM (1982) Sympathetic nerves protect against blood-brain barrier disruption in the spontaneously hypertensive rat. Stroke 13:83–88
Mueller SM, Ertel PJ, Felten DL, Overhage JM (1983) The chronic influence of sympathetic nerves on cerebral vessels is age-related. Stroke 14:286–289
Muramatsu I, Kushima SI, Fujiwara M (1977) Enhancement by the phentolamine of sympathetic contraction of dog basilar artery and alpha-receptor mediated feedback. In: Owman Ch, Edvinsson L (eds) Neurogenic control of the brain circula-

tion. Werner-Gren Center international symposium series, vol 30. Pergamon, Oxford, p 197
Nakai M, Iadecola C, Reis DJ (1982) Global cerebral vasodilation by stimulation of rat fastigial cerebellar nucleus. Am J Physiol 243:H226–H235
Nakai M, Iadecola C, Ruggiero DA, Tucker LW, Reis DJ (1983) Electrical stimulation of cerebellar fastigial nucleus increases cerebral cortical blood flow without change in local metabolism: Evidence for an intrinsic system in brain for primary vasodilation. Brain Res 260:35–49
Napolitano LM, Williams VL, Hanlon CR, Cooper T (1965) Intrinsic innervation of the heart. Am J Physiol 208:455–458
Nielsen KC, Owman CH (1967) Adrenergic innervation of pial arteries related to the circle of Willis in the cat. Brain Res 6:773–776
Nilsson B, Nordberg K, Nordström CH, Siesjö BK (1975) Influence of hypoxia and hypercapnia in rats. In: Harper M et al. (eds) Blood flow and metabolism in brain. Churchill Livingstone, Edinburgh, pp 9.19–9.23
Nilsson B, Rehncrona S, Siesjö BK (1979) Coupling of cerebral metabolism and blood flow in epileptic seizures, hypoxia and hypoglycaemia. In: Ciba Symposium on Cerebral Vascular Smooth Muscle and its Control. Excerpta Medica, Amsterdam, pp 199–214
Nordberg K, Siesjö BK (1975) Cerebral metabolism in hypoxic hypoxia. I. Pattern of activation of glycolysis: a re-evaluation. Brain Res 86:31–44
Nornes H, Knutgen HB, Wikeby P (1977) Cerebral arterial blood flow and aneurysm surgery. J Neurosurg 47:819–827
Oldendorf WH (1971) Brain uptake of radiolabeled amino acids, amines, and hexoses after arterial injection. Am J Physiol 221:1629–1639
Olesen J (1971) Contralateral focal increase of regional cerebral blood flow in man during arm work. Brain 94:635–646
Olsson RA, Saito D, Steinhart CR (1982) Compartmentalization of the adenosine pool of dog and rat hearts. Circ Res 50:617–626
Orkland RK, Nicholls JG, Kuffler SW (1966) Effects of nerve impulses on the membrane potential of glial cells in the central nervous system of amphibia. J Neurophysiol 29:788–806
Orr JA, Wagerle LC, Kiorpes AL, Shirer HW, Friesen BS (1983) Distribution of internal carotid artery blood flow in the pony. Am J Physiol 244:H142–H149
Pearce WJ, Bevan RD, Bevan JA (1982) Neurogenic vasoconstriction of extracranial veins draining canine cerebral venous effluent. In: Heistad DD, Marcus ML (eds) Cerebral blood flow: effects of nerves and neurotransmitters. Elsevier North Holland, New York, pp 87–84
Peerless SJ, Kendall MJ (1975) In: Smith RR, Robertson JT (eds) Subarachnoid hemorrhage and cerebrovascular spasm. Thomas, Springfield, pp 38–54
Peerless SJ, Yasargil MG (1971) Adrenergic innervation of the cerebral blood vessels in the rabbit. J Neurosurg 35:148–154
Pickard JD, MacKenzie ET (1973) Inhibition of prostaglandin synthesis and the response of baboon cerebral circulation to carbon dioxide. Nature New Biol 245: 187–188
Pickard DD, Rose JE, Cooke MBD, Blair IM, Strathdee A (1977) The effect of salicylate on cerebral blood flow in man. Acta Neurol Scand 56 (Suppl 64):422–423 (Abstract)
Pinard E, Purves MJ, Seylaz J, Vasquez JV (1979) The cholinergic pathway to cerebral blood vessels. II. Physiological Studies. Pflügers Arch 379:165–172
Ponte J, Purves MJ (1974) The role of the carotid body chemoreceptors and carotid sinus baroreceptors in the control of cerebral blood vessels. J Physiol (Lond) 237: 315–340
Prince DA, Lux HD, Neher E (1973) Measurement of extracellular potassium activity in cat cortex. Brain Res 50:489–495

Pull I, McIlwain H (1972) Metabolism of [^{14}C] adenine and derivates by cerebral tissues, superfused and electrically stimulated. Biochem J 126:965–973

Purdy RE, Bevan JA (1977) Adrenergic innervation of large cerebral blood vessels of the rabbit studied by fluorescence microscopy: Absence of features that might contribute to non-uniform change in cerebral blood flow. Stroke 8:82–87

Purves MJ (1978) Do vasomotor nerves significantly regulate cerebral blood flow? Circ Res 43:485–493

Raichle ME, Hartman BK, Eichling JO, Sharpe LG (1975) Central noradrenergic regulation of cerebral blood flow and vascular permeability. Proc Natl Acad Sci USA 72:3726–3730

Raichle ME, Grubb RL, Gado MH, Eichling JO, Ter-Pogossian MM (1976) Correlation between regional cerebral blood flow and oxidative metabolism. Arch Neurol 33: 523–526

Rapela CE, Green HD (1964) Autoregulation of canine cerebral blood flow. Circ Res 14 (Suppl 1):205–211

Rapela CE, Green HD, Denison AB Jr (1967) Baroreceptor reflexes and autoregulation of cerebral blood flow in the dog. Circ Res 21:559–568

Raper AJ, Kontos HA, Patterson JL Jr (1971) Response of pial precapillary vessels to changes in arterial carbon dioxide tension. Circ Res 28:518–523

Rapoport SI (1976) Blood-brain barrier in physiology and medicine. Raven, New York, pp 66–70

Rennels MG, Forbes MS, Anders JJ, Nelson E (1977) Innervation of the microcirculation in the central nervous system and other tissues. In: Owman C, Edvinsson L (eds) Neurogenic control of the brain circulation. Pergamon, Oxford, pp 91–104

Rosenberg AA, Jones MD Jr, Traystman RJ, Simmons MA, Molteni RA (1982) Response of cerebral blood flow to changes in P_{CO_2} in fetal, newborn, and adult sheep. Am J Physiol 242:H862–H866

Rosenblum WI (1976a) The "richness" of sympathetic innervation. A comparison of cerebral blood vessels. Stroke 7:270–271

Rosenblum WI (1976b) Some physiologic properties of nerves in the adventitia of cerebral blood vessels as revealed by fluorescence microscopy. In: Cervós-Navarro J, Betz E, Matakas F, Wüllenweber R (eds) The cerebral vessel wall. Raven, New York, pp 183–189

Rovere AA, Scremin OU, Beresi MR, Raynald AC, Giardini A (1973) Cholinergic mechanism in the cerebrovascular action of carbon dioxide. Stroke 4:969–972

Rubio R, Berne RM, Bockman EL, Curnish RR (1975) Relationship between adenosine concentration and oxygen supply in rat brain. Am J Physiol 228:1896–1902

Sadoshima S, Fujishima M, Tamaki K, Nakatomi Y, Ishitsuka T, Ogata J, Omal T (1980) Response of cortical and pial arteries to changes of arterial CO_2 tension in rats – a morphometric study. Brain Res 189:115–120

Sadoshima S, Busija DW, Brody MJ, Heistad DD (1981a) Sympathetic nerves protect against stroke in stroke-prone hypertensive rats. A preliminary study. Hypertension 3:I124–I127

Sadoshima S, Thames M, Heistad D (1981b) Cerebral blood flow during elevation of intracranial pressure: Effects of sympathetic nerves. Am J Physiol 10:H78–H84

Sadoshima S, Busija DW, Heistad DD (1983) Mechanism of protection against stroke in stroke-prone spontaneously hypertensive rats. Am J Physiol 244:H406–H412

Sagawa K, Guyton AC (1961) Pressure-flow relationships in isolated canine cerebral circulation. Am J Physiol 200:711–714

Sakabe T, Siesjö BK (1979) The effect of indomethacin as the blood flow-metabolism couple in the brain under normal, hypercapnic and hypoxic conditions. Acta Physiol Scand 107:283–284

Sakurada O, Kennedy C, Jehle J, Brown JD, Carbin GL, Sokoloff L (1978) Measurement of local cerebral blood flow with [^{14}C]iodoantipyrine. Am J Physiol 234: H59–H66

Schneider W, Wahl M, Kischinsky W, Thurau K (1977) The use of microelectrodes for measurement of local H^+ activity in the cortical subarachnoid space of cats. Pflügers Arch 372:103–107

Schrader J (1983) Metabolism of adenosine and sites of production in the heart. In: Berne RM, Rall TW, Rubio R (eds) Regulatory function of adenosine. Martinus Nijhoff, The Hague, pp 133–156

Schuier FJ, Fedora T, Jones SC, Reivich M (1981) Comparison of rCBF obtained by the microsphere method versus the C14-iodoantipyrine method. J Cereb Blood Flow Metab 1 (Suppl 1):S76–S77 (Abstract)

Schultz V, Lowenstein JM (1978) The purine nucleotide cycle. J Biol Chem 253: 1938–1943

Scremin OU, Rubinstein EH, Sonnenschein RR (1978) Cerebrovascular CO_2 reactivity: role of a cholinergic mechanism modulated by anesthesia. Stroke 9:160–165

Scremin OU, Sonnenschein RR, Rubinstein EH (1983) Cholinergic cerebral vasodilatation: lack of involvement of cranial parasympathetic nerves. J Cereb Blood Flow Metab 3:362–368

Sercombe R, Lacombe P, Aubineau P, Mamo H, Pinard E, Reynier-Rebuffel AM, Seylaz J (1979) Is there an active mechanism limiting the influence of the sympathetic system on the cerebral vascular bed? Evidence for vasomotor escape from sympathetic stimulation in the rabbit. Brain Res 164:81–102

Shalit MN, Reinmuth OM, Shimojyo S, Scheinberg P (1967) Carbon dioxide and cerebral circulatory control. III. The effects of brain stem lesions. Arch Neurol 17:342–353

Shepherd JT, Vanhoutte PM (1981) Local modulation of aderenergic neurotransmission. Circulation 64:655–666

Siesjö BK, Zwetnow NM (1970) The effect of hypovolemic hypotension on extra- and intracellular acid-base parameters and energy metabolites in the rat brain. Acta Physiol Scand 79:114–124

Silver IA (1973) Local PO_2 in relation to intracellular pH, cell membrane potential and potassium leakage in hypoxia and shock. Adv Exp Med Biol 37A:223–232

Silver IA (1978) Cellular microenvironment in relation to local blood flow. Ciba Found Symp 56:49–61

Singer W, Lux HD (1973) Presynaptic depolarization and extracellular potassium in the cat lateral geniculate nucleus. Brain Res 64:17–33

Singer W, Lux HD (1975) Extracellular potassium gradients and visual receptive fields in the cat striate cortex. Brain Res 96:378–383

Skinhøj E (1966) Regulation of cerebral blood flow as a single function of the interstitial pH in the brain. A hypothesis. Acta Neurol Scand 42:604–607

Skinhøj E, Strandgaard S (1973) Pathogenesis of hypertensive encephalopathy. Lancet 1:461–462

Sokoloff L (1981) Relationships among local functional activity, energy metabolism, and blood flow in the central nervous system. Fed Proc 40:2311–2316

Sokoloff L, Reivich M, Kennedy D, Des Rosiers CS, Patlak KD, Pettigrew KD, Sakurada O, Shinohara M (1977) The [^{14}C]deoxyglucose method for the measurement of local cerebral glucose utilization: Theory, procedure, and normal values for the conscious and anesthetized albino rat. J Neurochem 28:897–916

Solberg LA, Egger DA (1971) Localization and sequence of development of atherosclerotic lesions in the carotid and vertebral arteries. Circulation 43:711–724

Stehbens WE (1972) Pathology of cerebral blood vessels. Mosby, St Louis, pp 98–130

Strandgaard S (1976) Autoregulation of cerebral blood flow in hypertensive patients. The modifying influence of prolonged antihypertensive treatment on the tolerance to acute, drug-induced hypotension. Circulation 53:720–727

Strandgaard S, Olesen J, Skinhoj E, Lassen NA (1973) Autoregulation of brain circulation in severe arterial hypertension. Br Med J 1:507–510

Strandgaard S, MacKenzie ET, Sengupta D, Rowan JO, Lasser NA, Harper AM (1974) Upper limit of autoregulation of cerebral blood flow in the baboon. Circ Res 34: 435–440

Strandgaard S, Jones JV, MacKenzie ET, Harper AM (1975) Upper limit of cerebral blood flow autoregulation in experimental renovascular hypertension in the baboon. Circ Res 37:164–167

Suzuki H, Fujiwara S (1982) Neurogenic electrical responses of single smooth muscle cells of the dog middle cerebral artery. Circ Res 51:751–759

Symon L, Held K, Doroch NWC (1973) A study of regional autoregulation in the cerebral circulation to increased perfusion pressure in normocapnia and hypercapnia. Stroke 4:139–147

Toda N (1982) Relaxant responses to transmural stimulation and nicotine of dog and monkey cerebral arteries. Am J Physiol 243:H145–H153

Toda N, Fujita Y (1973) Responsiveness of isolated cerebral and peripheral arteries to serotonin, norepinephrine and transmural electrical stimulation. Circ Res 33:98–104

Toda N, Hatano Y, Hayaski S (1978) Modifications by stretches of the mechanical response of isolated cerebral and extracerebral arteries to vasocative agents. Pflügers Arch 374:73–77

Towart R (1981) The selective inhibition of serotonin-induced contractions of rabbit cerebral vascular smooth muscle by calcium-antagonistic dihydropyridines. An investigation of the mechanism of action of nimodipine. Circ Res 48:650–657

Towart R, Kazda S (1982) Preferential vasodilator actions of the calcium antagonists nimodipine (BAYe 9736), nifedipine, and verapamil on contractions of cerebral vascular smooth muscle induced by neurotransmitter and vasoconstrictor substances. In: Heistad DD, Marcus ML (eds) Cerebral blood flow: effects of nerves and neurotransmitters. Elsevier North Holland, New York, pp 29–38

Traystman RJ, Fitzgerald RS (1981) Cerebrovascular response to hypoxia in baroreceptor- and chemoreceptor-denervated dogs. Am J Physiol 241:H724–H731

Traystman RJ, Rapela CE (1975) Effect of sympathetic nerve stimulation on cerebral and cephalic blood flow in dogs. Circ Res 36:620–630

Urbanics R, Leniger-Follert E, Lübbers DW (1978) Time course of changes of extracellular H^+ and K^+ activities during and after direct electrical stimulation of the brain cortex. Pflügers Arch 378:47–53

Vasquez J, Purves MJ (1979) The cholinergic pathway to cerebral blood vessels. I Morphological Studies. Pflügers Arch 379:157–163

Vatner SF, Priano LL, Rutherford JD, Manders WT (1980) Sympathetic regulation of the cerebral circulation by the carotid chemoreceptor reflex. Am J Physiol 238: H594–H598

Vindall PE, Simeone FA (1982) In vitro myogenic autoregulation in cerebral blood vessels. In: Heistad DD, Marcus ML (eds) Cerebral blood flow. Effect of nerves and neurotransmitters. Elsevier, New York, pp 57–66

Wagerle LC, Heffernan TM, Sacks LM, Delivoria-Papadopoulos M (1983) Sympathetic effect on cerebral blood flow regulation in hypoxic newborn lambs. Am J Physiol 245:H487–H494

Wagner EM (1982) Hydrostatic determinants of cerebral blood flow. Ph D Dissertation, Johns Hopkins University, Baltimore

Wagner EM, Traystman RJ (1983) Cerebral venous outflow and arterial microsphere flow with elevated venous pressure. Am J Physiol 244:H505–H512

Wahl M, Kuschinsky W (1976) The dilatory action of adenosine on pial arteries of cats and its inhibition by theophylline. Pflügers Arch 362:55–59

Wahl M, Kuschinsky W (1979) Unimportance of perivascular H^+ and K^+ activities for the adjustment of pial arterial diameter during changes of arterial blood pressure in cats. Pflügers Arch 382:203–208

Wahl M, Kuschinsky W, Bosse O, Thurau K (1973) Dependency of pial arterial and arteriolar diameter on perivascular osmolarity in the cat. Circ Res 32:162–169

Wei EP, Kontos HA (1982) Responses of cerebral arterioles to increased venous pressure. Am J Physiol 243:H442–H447

Wei EP, Kontos HA (1983) Role of adenosine in cerebral arteriolar dilation from arteriolar dilation from arterial hypoxia. J Cereb Blood Flow Metab 1 (Suppl 1): S95–S96

Wei EP, Raper AJ, Kontos HA, Patterson JL Jr (1975) Determinants of response of pial arteries to norepinephrine and sympathetic nerve stimulation. Stroke 6:654–658

Wei EP, Dietrich WD, Povlishock JT, Navari RM, Kontos HA (1980a) Functional, morphological, and metabolic abnormalities of the cerebral microcirculation after concussive brain injury in cats. Circ Res 46:37–47

Wei EP, Ellis EF, Kontos HA (1980b) Role of prostaglandins in pial arteriolar response to CO_2 and hypoxia. Am J Physiol 238:H226–H230

Wei EP, Kontos HA, Patterson JL Jr (1980c) Dependence of pial arteriolar response to hypercapnia on vessel size. Am J Physiol 238:H697–H703

Wei EP, Kontos HA, Said SI (1980d) Mechanism of action of vasoactive intestinal polypeptide on cerebral arterioles. Am J Physiol 239:H765–H768

Weiss HR, Buckweitz E, Sinha AK (1983) Effect of hypoxic-hypocapnia on cerebral regional oxygen consumption and supply. Microvasc Res 25:194–204

Wellens D, Wouters L, Stijner L, Reneman RS (1977) Anatomical and pharmacological characteristics of the cerebral circulation in dogs, cats and rats: contribution of the radioactive microsphere method. Bibl Anat 15:82–86

Wilson DA, O'Neill JT, Said SI, Traystman RJ (1981) Vasoactive intestinal polypeptide and the canine cerebral circulation. Circ Res 48:138–148

Winquist RJ, Webb RC, Bohr DF (1982) Relaxation of transmural nerve stimulation and exogenously added norepinephrine in porcine cerebral vessels: a study utilizing cerebrovascular intrinsic tone. Circ Res 51:769–776

Winn HR, Rubio R, Berne RM (1979) Brain adenosine production in the rat during 60 seconds of ischemia. Circ Res 45:486–492

Winn HR, Welsh JE, Rubio R, Berne RM (1980a) Brain adenosine production in rat during sustained alteration in systemic blood pressure. Am J Physiol 239:H636–H641

Winn HR, Welsh JE, Rubio R, Berne RM (1980b) Changes in brain adenosine during bicuculline-induced seizures in rats. Circ Res 47:568–577

Winn HR, Rubio R, Berne RM (1981) Brain adenosine concentration during hypoxia in rat. Am J Physiol 241:H235–H242

Author Index

Subject Index

Of Further Interest from this Series

Volume 93

1982. 22 figures. V, 213 pages. ISBN 3-540-11297-9

Contents: *H.-H. Wellhöner:* Tetanus Neurotoxin. – *R. D. Manning, Jr., A. C. Guyton:* Control of Blood Volume. – *P. C. Heinrich:* Proteolytic Processing of Polypeptides During the Biosynthesis of Subcellular Structures.

Volume 94

1982. 52 figures. V, 207 pages. ISBN 3-540-11701-6

Contents: *M. Vogt:* Obituary Joshŭa Harold Burn (1892–1981). – *A. Malliani:* Cardiovascular Sympathetic Afferent Fibers. – *G. F. DiBona:* The Functions of the Renal Nerves.

Volume 95

1983. 50 figures. V, 223 pages. ISBN 3-540-11736-9

Contents: *V. B. Brooks:* Study of Brain Functions by Local, Reversible Cooling. – *H. Rasmussen, D. M. Waisman:* Modulation of Cell Function in the Calcium Messenger System. – *K. E. O. Åkerman, D. G. Nicholls:* Physiological and Bioenergetic Aspects of Mitochondrial Calcium Transport.

Volume 96

1983. 21 figures. V, 194 pages. ISBN 3-540-11849-7

Contents: *H. Murer, G. Burckhardt:* Membrane Transport of Anions Across Epithelia of Mammalian Small Intestine and Kidney Proximal Tubule. – *W. Osswald, S. Guimarães:* Andrenergic Mechanisms in Blood Vessels: Morphological and Pharmacological Aspects. – *O. H. Wieland:* The Mammalian Pyruvate Dehydrogenase Complex: Structure and Regulation.

Volume 97

1983. 15 figures. V, 176 pages. ISBN 3-540-12135-8

Contents: *P. R. Stanfield:* Tetraethylammonium Ions and the Potassium Permeability of Excitable Cells. – *G. A. Quamme, J. H. Dirks:* Renal Magnesium Transport. – *R. Fänge:* Gas Exchange in Fish Swim Bladder.

Volume 98

1983. 94 figures. V, 255 pages. ISBN 3-540-12817-4

Contents: *H. Blaschko:* Obituary Hans Adolf Krebs 1900–1981. – *J. Szentágothai:* The Modular Architectonic Principle of Neural Centers. – *S. D. Erulkar:* The Modulation of Neurotransmitter Release at Synaptic Junctions. – *J. Krüger:* Simultaneous Individual Recordings From Many Cerebral Neurons: Techniques and Results.

Volume 99

1984. 36 figures. V, 227 pages. ISBN 3-540-12989-8

Contents: *J. C. G. Coleridge:* Afferent Vagal C Fibre Innervation of the Lungs and Airways and Its Functional Significance. – *H. Pfister:* Biology and Biochemistry of Papillomaviruses. – *G. G. Pinter, K. Gärtner:* Peritubular Capillary, Interstitium, and Lymph of the Renal Cortex.

Volume 100

1984. 20 figures. V, 247 pages. ISBN 3-540-13327-5

Contents: *J. Maj, E. Przegaliński, E. Mogilnicka:* Hypotheses Concerning the Mechanism of Action of Antidepressant Drugs. – *K. Pleschka:* Control of Tongue Blood Flow in Regulation of Heat Loss in Mammals. – *T. Ruzicka, M. P. Printz:* Arachidonic Acid Metabolism in Skin: A Review. – *J.-P. Bonjour, J. Caverzasio:* Phosphate Transport in the Kidney. – Author Index. – Subject Index.

Springer-Verlag
Berlin
Heidelberg
New York
Tokyo

GPSR Compliance

The European Union's (EU) General Product Safety Regulation (GPSR) is a set of rules that requires consumer products to be safe and our obligations to ensure this.

If you have any concerns about our products, you can contact us on ProductSafety@springernature.com

In case Publisher is established outside the EU, the EU authorized representative is:

Springer Nature Customer Service Center GmbH
Europaplatz 3
69115 Heidelberg, Germany

Batch number: 10620881

Printed by Printforce, the Netherlands